Valerie Horne
July 2004
17-18 Seminar

Entities

Parasites of the Body of Energy

By the same author:

✳ **Awakening the Third Eye**

To be released in 1994:

✳ **Regression, Past-Life Therapy for Here and Now Freedom**

In preparation:

✳ **Planetary Forces, Alchemy and Medicine**

✳ **The Moon Cycle**

✳ **Psychic Experiences and How to Handle Them**

✳ **Tantra, Body and Worlds**

The Clairvision School also offers a range of correspondence courses dealing with subtle bodies, esotericism, spiritual development and astrology. Available modules include:

✳ **An Introduction to Subtle Bodies**

✳ **Canopus (a computer software to learn astrology)**

Clairvision School

The Clairvision School deals with spiritual development in line with the western tradition of esoteric knowledge. The approach of the school is resolutely experiential. It is designed for people who cannot be satisfied with intellectual understanding only and who wish to gain a first-hand experience of spiritual realities.

In essence, Clairvision stands for the higher vision of the Self. As such, it has little to do with the clairvoyant flashes of psychic people. Clairvision is not a passive opening to random experiences but an active function by which the Self knows Itself and the creation — as in the adage engraved on the portal of Delphi's oracle, "Know Thyself and Thou Shalt Know the Universe and All the Gods".

You are the alchemist, your bodies (subtle and physical) are the laboratory. The Clairvision School's program is an inner journey through your own spheres of consciousness. It uses a rich arsenal of processes and techniques for discovering the various facets of your energy and linking you to your Ego—capital E.

The essential purpose of the Clairvision School is to offer an experiential work of quality and to train people to a high level in the fields of self-transformation, spiritual development and esoteric knowledge. The courses of the school are therefore intense, and are designed for people who are strongly motivated and sincere in their approach. In order to join, no prior training or knowledge is required, just an open mind and a real aspiration to transform oneself.

Apart from long term weekly courses conducted in Sydney and other places in Australia, the school runs intensive residential courses especially designed for interstate and overseas students. These international courses include a thorough training in meditation, development of the chakras and circulations of energy, and ISIS, the school's technique of self-exploration and regression. More advanced aspects are related to inner alchemy and the building of the subtle bodies, the transformation of thinking, and knowledge of spiritual beings.

Clairvision School, PO Box 33, Roseville NSW 2069, Australia

Entities

Parasites of the Body of Energy

Dr. Samuel Sagan

Cover and illustrations by
Steve Goldsmith

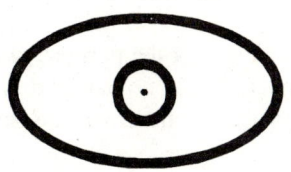

Clairbision School
PO Box 33, Roseville NSW 2069, Australia

Copyright © 1994 by Clairvision School Foundation
Published in Sydney, Australia, by Clairvision School
PO Box 33, Roseville NSW 2069, Australia

Apart from any fair dealing for the purpose of review or research, as permitted under the Copyright Act, no part of this book may be reproduced by any process without written permission.

ISBN 0-646-17882-2

Clairvision is a trademark of Clairvision School Ltd.

Thanks to the actor Mattheus Levell
who posed for the cover of this book.

Table of Contents

Introduction 1
Chapter 1 – The facts 4
Chapter 2 – The facts (continued) 15
Chapter 3 – Taoist perspectives on entities 30
Chapter 4 – Fragments 40
Chapter 5 – Fragments (continued) 52
Chapter 6 – Entities, pregnancy and gynaecology 63
Chapter 7 – Entities, pregnancy and gynaecology (continued) 75
Chapter 8 – How does one catch an entity? 85
Chapter 9 – Entities of various kinds 98
Chapter 10 – Entities and past lives 107
Chapter 11 – Cords 120
Chapter 12 – Missing bits 133
Chapter 13 – Possession and extraordinary entities 138
Chapter 14 – Exploring an entity 152
Chapter 15 – Clearing entities 165
Conclusion 179
Index 180

INTRODUCTION

The term 'entity' refers to non-physical beings, presences which come to be attached to human beings and act as parasites, thereby creating various emotional, mental and physical problems ranging from eating disorders and uncontrollable emotions to the most severe diseases.

The topic is both old and new. Old, because in all traditions and folklores of the earth, one finds references to spirits and non-physical beings which can interfere with human beings. Thus Ayurveda, the traditional medicine of India, is divided into eight sections, one of which is entirely devoted to the study of *bhūtas*, or entities, their influence on health and sanity, and the ways one can get rid of them. This places *bhūta-vidyā*, or 'science of entities', on the same level as surgery or gynaecology. If we look at traditional Chinese medicine, we find that in acupuncture, among the 361 points of the 14 main meridians, 17 have the word *Kuei* (disincarnate spirit) as part of their main or secondary name.

At the same time, in the western world, the topic of entities can be regarded as quite new, for it has very rarely been studied seriously. Even though entities appear to play a significant role in a number of mental and health disorders, minor and major, one does not find a systematic way of dealing with them in any of the main forms of therapy presently used. The number of qualified therapists capable of handling entities properly is negligible.

The purpose of this book is to present certain basic facts relating to entities and to analyse their mechanisms, motivations and functioning. Over the years, I have noted word for word what my clients have said when getting in touch with entities and exploring them. It should be emphasised that these clients comprised people of all ages, from teenage to late eighties, and with varied occupations, from housewives and students to politicians, prostitutes and Catholic nuns. Perhaps the only thing these people had in common was that none had the faintest knowledge of entities and their mechanisms. Moreover, I *never* told them at the beginning of the session that I considered they had an entity. I let them discover it for themselves through the ISIS technique (the main principles of which are outlined below). Yet when analysing the material contained in hundreds of case studies and related observations, a remarkably coherent pattern emerged as to what entities want, how they interfere with their hosts, and why and how they came to invade them. I was astonished at the precision, the richness of detail and the consistency with which clients described their entity.

More than anything else it is this pattern, or 'entity syndrome', that I wish to share with the reader. In the first two chapters, I will present 'the facts' with the words used by the clients themselves while discovering and exploring their entity.

Chapters 3, 4 and 5 describe a pattern aimed at understanding what most entities are, where they originated, and the 'why and how' of their behaviour.

Chapters 6 and 7 are devoted to the extremely important topic of the influence of entities in the fields of pregnancy and gynaecology. One of my findings has been that delivery, miscarriage and abortion are among the high-risk situations for catching an entity, and some basic information on the topic can save women a lot of problems.

Chapter 8 describes the main circumstances in which it is possible to catch an entity. Chapter 9 presents an inventory of the main categories of beings that can act as entities.

In Chapters 10, 11 and 12, a number of side-mechanisms related to entities are described, in particular cords of energy linking parents and children, thereby creating distorted patterns of relationship.

Chapter 13 discusses the topic of possession and presents a few observations of 'extraordinary entities'. Yet it should be clear that, while treating clients with entities, I have found the cliches related to exorcism and possession irrelevant, unfortunate, and counterproductive to a sound approach to the phenomenon. It is time for the public to start facing the topic of entities squarely, instead of being negatively influenced and misled by all the folklore attached to it. The purpose of this book is therefore to de-dramatise and demystify the topic, by looking at entities from the point of view of experience and mechanisms, rather than from religious or emotionally tainted perspectives.

In Chapters 14 and 15, indications are given on the procedure we use to explore and remove entities at the Clairvision School in Sydney. A number of restrictions and possible dangers related to the clearing process will be dealt with in detail.

Introduction

One of the observations I have made in recent years is that more and more people are able to perceive entities. It therefore seems likely to me that the topic treated in this book will, in the coming decades, concern an ever growing number of therapists and spiritual seekers.

                                      Clairvision School
                                      PO Box 33
                                      Roseville NSW 2069
                                                Australia

CHAPTER 1

THE FACTS

1.1 ISIS

A powerful technique of inner exploration, called ISIS, will be mentioned frequently throughout this book. The purpose of ISIS is to develop inner vision by unveiling the sources of emotional blockages and psychological dispositions. Through ISIS, the roots of conditioned behaviour are explored and released, to enable us to reach a clarity and a spontaneity of a higher level.

The ISIS technique is based on three main principles: Interaction, Inner Space and Sourcing.

✧**Interaction**—ISIS is practised on a one-to-one basis. The client lies down with closed eyes while the facilitator, called 'connector', sits nearby. The interaction that takes place between the energy of the client and that of the connector is an essential part of the process. In particular, at times during the practice the connector asks the client certain questions to help reveal significant patterns in the client's psyche.

✧**Inner Space**—through a particular method of activating the energy centre between the eyebrows, clients are put in touch with the inner space of consciousness. They become aware of a purple light or purple expanse, and an opening of perception often takes place, in which a number of emotional energies can be discerned.

✧**Sourcing**—the general direction of the technique is to locate the source of the emotions and energies met while scanning the inner space.

The initials of these three principles propitiously make up the name of the Egyptian Goddess Isis.

I will not dwell on the procedural details of ISIS as these have been amply described elsewhere.[1] However, I must emphasise that the process

---

[1] See *Regression, Past Life Therapy for Here and Now Freedom* by the same author.

does not involve any form of hypnosis, suggestion, creative visualisation, guided imagination or positive affirmation. Instead, clients are encouraged to look at the inner images in a non-manipulative way, not trying to modify anything of what they see. Moreover, they remain fully conscious and awake for the entire duration of the session. Whilst in the ISIS state, an expansion of perception takes place, and clients become aware of a number of movements and energies inside themselves that they had never been able to discern before.

1.2 The facts

In the first two chapters, I will present the 'facts' — observations made by the clients themselves while discovering an entity through the ISIS technique. It should also be emphasised that I *never* tell clients if I perceive that they have an entity. The direction of the Clairvision School's techniques is to develop free will, and let clients and students play as active a role as possible in any process of healing or self-transformation. Therefore, in ISIS, the connector never declares there is an entity before the clients themselves have perceived it.

Let us now go through the most common findings made by clients while discovering and exploring an entity with the ISIS technique.

1.3 Presence

In most cases, clients can feel a presence attached to them. In other words, they perceive an autonomous consciousness, added to their own and separate from it, and operating within or around them.

Most clients are unaware of that presence before undergoing the exploration process. However, once identified through ISIS, the presence often feels familiar to them. They suddenly realise that they already knew about it subconsciously.

A certain percentage of clients were consciously aware of the presence long before undergoing the ISIS process, but were unable to talk about it with friends or therapists, for fear of being regarded as insane, or just because it sounded peculiar and did not fit in with any of their usual mental concepts.

1.4 Separateness

In almost all cases, clients perceive the presence as separate from themselves. They can feel their own presence plus another presence, which is attached to them but distinct from them. They describe it in the following terms: "something foreign", "it doesn't belong to me", "it isn't part of me", "it is a parasite", "it hasn't always been with me", "it was added", "it got attached to me at a particular moment in time" (even if that time was long ago), "it feels wrong", "it shouldn't be there".

Case study 1.1 Thirty-two year old woman, secretary.

What can you perceive? –It's red, it's angry. It's as if I've been taken over. It's something that I've been fighting all my life. It inspires hatred. It's madness, but not madness from the body. Maybe it's my dark side. But it does not feel like me.
When did you feel it for the first time? –Now, actually. But I have always known it was there. [The client is crying, with her fists clenched.] I feel that it could make me kill, and I mean it. It fills up my body with hatred, it's sheer destruction.[1]

As we will see throughout this chapter, clients describe these foreign presences as having their own desires, emotions and thoughts. The presences are perceived as having an existence of their own, even though it interferes in many ways with the clients' psychology and vital functions.

1.5 Aspect, location and size

In most cases, clients come to identify a shape associated with the presence. These shapes vary from simple shadows to human shapes and a wide range of monstrous forms. Even though clients may not always verbalise it clearly, they do perceive the shape as the vehicle of the presence, just as our physical body can be regarded as the vehicle of our consciousness.

The shape is usually first identified in the ISIS state. Later on, during their daily activities, most clients can remain aware of the presence attached to them. When asked whether this shape was already attached to

---

[1] The case studies presented in this book are in the form of a dialogue between the connector (facilitator), who asks questions, and the client who answers them or makes remarks of his/her own. All these dialogues have taken place while the clients were in the ISIS state. The clients' exact words have been kept, apart from very minor adjustments of syntax and the elimination of a few repetitions. In order to recognise who is speaking, the facilitator's words are written in italics, and a dash (–) is used when the client starts speaking.

The Facts

them before seeing it in ISIS, clients almost invariably answer "yes", but that they were unaware of it, or only aware of it subconsciously.

The identification of the shape is often gradual. For instance, in the beginning of the session the client may just see a blurry cloud, and only gradually perceive more and more details until the full shape is revealed. Once fully identified, the shape usually remains roughly the same throughout the exploration process, until the clearing. However, the size may vary slightly. In particular, certain emotions or foods tend to make the entity temporarily bigger.

In most cases, clients describe these forms as being inside their body or directly attached to it, for instance stuck to their back or sitting on top of their head. It follows that the shape/presence is located in one particular part of their body. The location most often identified by both men and women is the left iliac area. Women also frequently identify areas in and around the vagina, uterus and ovaries. The reasons why this may be so are discussed in Chapter 6. For the moment, let us focus on the facts as presented by the clients.

Entities can be located in virtually any body part, but the limbs come up quite rarely compared to the trunk and the head. An important finding is that entities do not seem to move around much in the client's body. Sometimes clients relate that 'the thing' moves within its own space slightly, but it is unusual for it to change body area altogether. Entities appear very fixed and stubborn, unwilling to let go of their position, and unresponsive to any attempt to dislodge them.

The size of entities, as described by clients, varies from half a centimetre to two metres. However, in the majority of cases, it is less than 50 centimetres.

1.6 Draining the client's life force

In nearly all cases, clients report that 'the thing' is draining them, tapping from their life force. This is quite a constant feature, as will be seen in examples throughout the book.

Case study 1.2 Forty-three year old woman, air hostess.

What does 'it' look like? –It's got legs with suckers at the end. The two front legs are around my neck. It's big. It covers my back, down to the middle of my back. It's grey. It feeds off the base of my brain. It takes life energy but not all the time, just sometimes, when it's hungry. When it

sucks my energy, I get confused. It gives round and round thoughts, thoughts that do not make any sense. It bites my head and my head hurts.
Does it feel like a part of yourself or like something foreign? –No, it's foreign.

Case study 1.3 Twenty-nine year old woman, housewife. She suffered from fatigue without any reason, depression and lack of motivation. At the beginning of the ISIS session, the client perceived a shadow in her left iliac area.

Can any emotion be related to it? –It just feels dead. Cold. There's nothing really. It just likes to be left alone.
Why? –It likes the dark. Like in a cave. It likes to be still and to draw things into itself.
How big is it? –You mean that black thing... [The client indicates an area from mid-thigh to her breast, on the left side of her body.]
Have you seen it before? –No.
What does it want? –It just wants to be there. It takes little bits of my energy. It sucks energy all the time. Just little bits. Just what's necessary.
Could it be that there are some foods you eat that it enjoys? –Cheese. And tomatoes. Bread.
What happens to it when you have these foods? –It gets stronger. And my energy gets a bit weaker.
What emotions of yours does it enjoy? –Negativity. It's like a pattern of negativity. Empty life type of thing. Depression. It likes my husband. My husband feeds it. He makes it stronger.

Soon after identifying the presence, most clients tend to perceive it as being a parasite. In order to describe it, they use expressions such as: "It's draining me", "It lives on my energy", "It taps from my life force", "It's sucking my energy", "It takes my life (or my warmth)", "It feeds on me", "I'm its life-support system". They often relate the presence to symptoms of fatigue, emptiness, depression and lack of motivation.

1.7 Food cravings

In most cases, clients report that when they eat certain foods, the presence reacts. Moreover, the presence can create cravings or compulsive desires for these particular foods.

The Facts

The substance that is mentioned by far the most often is sugar. When clients are asked "Could it be that there are some foods that 'the thing' enjoys?", more than half of them immediately answer "sugar", "sweets" or "chocolate".

Other foods often mentioned by clients are 'heavy foods', cakes and white bread (meaning yeast), fried foods and fats, cheese, and junk food in general. In other cases, meat and spicy or salty foods are preferred. It is not unusual for meat and spicy foods to be mentioned together. A pattern I have observed is that the entities which have cravings for meat are often more attracted to spices, wine and alcohol than to cakes and sweets, and are frequently associated with angry and aggressive tendencies.

Surprisingly, one vegetable is often reported by clients among the 'entity cravings' — tomatoes.

It should be stressed that as soon as the client has identified the presence during a session, he or she usually finds it very easy to answer the question: "Could it be that there are some foods you eat that it enjoys?" The client's response tends to indicate that food cravings are one of the most obvious features related to entities. Not all presences have food cravings, but most of them do; and when they do, the cravings seem quite easy to identify.

Case study 1.4 Thirty-seven year old woman, film maker.

What does it look like? —Like a stone, in my solar plexus. It's quite old. It's not giving, and quite unforgiving.
Could it be that there are some foods you eat that it enjoys? —It likes meat and hot foods, like red chillies, and red wine. It's all red: red wine, red meat, red chillies, red capsicums, tomatoes... how bizarre! It's very aggressive. It's got nasty teeth. It likes to be with me because I protect it. I make it comfortable and easy. It's like a sucker.

In dealing with an entity, the Clairvision approach consists of having one or more sessions devoted to identifying the presence. This is the first part of the process. Next, clients undergo a second phase, during which they watch this presence in their daily activities, in order to find out for themselves what interference in their behaviour may stem from it. During this observation phase which lasts two weeks or more, clients have the opportunity to study the mechanisms of the cravings. A common finding is that, when a craving arises, there seems to be an impulse coming from the presence. The craving starts in the presence and is insidiously communicated to the client. When clients are unaware and forget about the presence, the craving is perceived just like any craving of their own. However, if they remain vigilant and keep watching the presence, the same craving is seen coming straight from the presence. Clients often come to the conclusion that for years, the presence had been superimposing its own impulses upon their consciousness, even though they had been unaware of the mechanism until then.

Another important fact repeatedly reported by clients is that when they yield to a craving coming from the presence and eat the particular food, it seems to reinforce the presence. They all describe this effect in more or less the same terms: "The entity gets bigger" (or stronger), "It has more hold on me", "It can influence me more", "It gets out of control", and other similar phrases.

It must be emphasised that even if most entities generate cravings, there is absolutely no point in worrying whether you have an entity each time you have a craving! Just as a headache does not indicate a brain tumour, it would be absurd to consider that any symptom mentioned in this chapter may in itself indicate that you have an entity.

1.8 Other cravings

Apart from sugar and other foods, clients sometimes also report that the presence is responsible for other cravings: coffee, tobacco, alcohol and various other drugs. Only a small fraction of my entity-clients say that the presence craves alcohol or narcotic drugs. This may be because most of the people I see are inclined towards self-transformation work and have chosen a healthier life-style than that of the average population. Experience has led me to believe that if the same process were to be conducted with a group of alcoholics and drug addicts, it would probably reveal a surprisingly high frequency of entities. Yet this is not to suggest that all addictions are due to entities! An attitude that blamed entities for all human troubles would be just as childish as one based on a complete denial of the experience.

Whether cravings have to do with food or toxic substances, the question that arises is, how do the cravings linked to the presence of an entity differ from usual ones? Entity cravings are often more compelling and rigidly imprinted than other ones. They may lead clients to make comments such as: "It doesn't feel like me", "It does not feel like my usual desires and cravings". Once clients have identified the presence, they can perceive the cravings as coming straight from it. After this identification, they can differentiate between cravings, desires and attractions coming from themselves, and others coming from the presence. This requires dis-

The Facts

cernment and vigilance, for entities are clever at disguising their needs and remaining unnoticed, so that clients will interpret any cravings or desires as their own.

In many cases, the cravings linked to an entity disappear or at least significantly subside after the clearing process is carried out.

Case study 1.5 Twenty year old woman, student.

What are you feeling? —I can feel movement, like an animal fidgeting inside my solar plexus.
What kind of animal? —A small red ape. He is inside.
What is it doing there? —Jumping, moving...
What does it want? —He is agitated. He wants to get out.
Has it been there for a long time? —Oh, yes! I'm sure he was already there when I was a child. There is a feeling of constriction. It's very small: two or three inches.
What kind of food does the monkey like you to have? —He gets me agitated so I take sedatives. He wants to get out. He likes it when I have sedatives, and all sorts of drugs.
Have you ever seen that monkey before? —No. Though the feeling is familiar. He likes coffee. I drink a lot of coffee. Extreme runabout. He likes to go to coffee shops with lots of people and lots of music. And then go somewhere else, and then somewhere else... He does not feel dangerous, he feels very tricky. At home he makes me drink a lot of coffee, listen to loud music, and invite a lot of people over. If I can't do that he makes me panic. He hates it when I meditate. If I meditate he turns into a white spiral and spins.
Does the monkey talk? —No, but I do. He makes me talk all day and be restless during my work. He likes coffee — a lot. When I met my first boyfriend, I remember this new feeling, spinning, restless and agitated, and uncomfortable... leading to panic. My father was not long dead. It couldn't have been there before, because I was too serious: no opportunity. After my father died, I suddenly changed.
Was it happy when you had a boyfriend? —He is never happy. Always agitated. He is really in a bubble, hitting the edges. He wants to get out, or wants attention — very naughty. When I started taking drugs [cocaine] he got much worse. He gets a lot wider and bigger and out of control [indicating the whole of her chest and abdomen with her hands].
Could there be a part of yourself that gets some benefit out of the presence of the monkey? —The social part. He plays on my darker part, and my

frivolous parts. I get a lot done with him. I can run a lot and see a lot of people, and stay awake at night. I can't read but I can talk, talk, talk... I can start to talk to a stranger sitting at the next table. Part of me likes that. But sometimes he is so irresistible that I have to run away from lectures at university, and light a cigarette. He loves Kings Cross [Sydney's red light district], he eats that energy, when I walk and talk to people.
How does it react when you have sex? –He withdraws. If I don't know the person or if there is more than one person, or a lot of alcohol, then he likes it. But anyone that I care for, he makes me panic and stops me from making love. I can see when it came in. It was a time... a sad time. Long ago. My father died when I was at school. I was running and I felt him die. I was thirteen. My brother and my uncle picked me up from school. And the ape jumped inside me in the panic that followed. My father used to take a lot of drugs.
–Now the ape is extremely angry. He hates you. He wants to hit your hand. He makes the whole of me tighten and try to push you away. He does not want me to be here. Each time he tries to convince me not to come here for the session.

1.9 Entities crave emotional intensity

As we have seen, clients describe how their entity creates cravings for sugar, junk foods or toxic substances, and then thrives on the enjoyment that follows. Similarly, most clients relate that the entity is reinforced when they experience any form of sensual enjoyment, or harsh emotions such as anger, dismay, frustration, any form of emotional pain, melancholy, sadness and depression.

Case study 1.6 Thirty-nine year old man, public servant.

–The bloody thing always wins. Either I have the cigarette and it thrives on the tobacco, or I don't and it thrives on my frustration!

Rather than the type of emotion, it is the emotion's intensity that seems to feed the entity. As far as sexual enjoyment is concerned, some entities appear to be especially attracted to sex, pushing the client to multiply experiences or partners. Others seem to be threatened by sexual intercourse and unable to cope with closeness and love. The general pattern is

The Facts

that entities seem to be more attracted to sex than to love, as was clearly expressed in case 1.5 involving the little red ape.

Case study 1.7 Twenty-four year old woman, nurse.

What are you feeling? —I can feel something in my chest and in my abdomen. It looks like an old man, like a skeleton. I can feel his ribcage superimposed on mine. I can feel his hip bones as well.

What does it want? —I think it wants me to be his lover. It likes me because I'm young. It likes my strength and my energy. It tries to camouflage things. It tries to make me feel my body, particularly my hips. It makes me walk in a certain way, remembering my hips sexually.

What does it get out of it? —It gets the feeling of being alive. And it likes this weather [it's spring in Sydney] because it's more sensual. It makes me feel the weather. There is a certain man, where I work... the skeleton is trying to manipulate me to have sex with him. Whenever I speak to that man, it modifies my voice to make it more attractive. And then in the evening it sends erotic scenes of me and that man into my mind. It pushes me to that man. It wants me to be touched by him. If the skeleton could get me to have sex with that man, it would get a sort of thrill out of it. Also it would get more power over me. It has set its mind on it, so if it achieves it it's like reinforcing its ego. And it would gain power from the sexual energy. It wants vaginal and anal sex. It would give it the feeling of extending itself to all of my body. Before I had my first boyfriend, it was making me feel unsure of myself. It was making me remember my body. And then I felt guilty and that gave it power over me. When I was fifteen I had my first boyfriend. I can see the skeleton was already there when I was making love, enjoying it lots. I remember a night when I was five or six. I had a dream about destruction and I had an orgasm and I woke up. It was the skeleton that made me have an orgasm. The skeleton was lying on my body, and the dream was so powerful that the skeleton had an orgasm.

Case study 1.8 Thirty-three year old woman, shopkeeper.

What does 'it' look like? —It has an amorphous shape, with extensions that end in sort of points. It's brownish dark. It goes from my hips to my shoulder, on my right side. It takes my life energy from there [indicating the area around her navel]. When I have fear, it can take more energy. The fear makes me weak, and it likes that. It makes it easier for it to grow strong. It

13

Entities

> wants to grow, until it can take me. If I had an accident, if I was dead, it could take over. It could control me, cause harm.

As we will soon see, it is common for entities to be stubbornly attached to one particular form of craving or desire.

CHAPTER 2

THE FACTS (continued)

2.1 Entities want something

Apart from their general tendency to thrive on emotional intensity and sensual enjoyment, most entities have a specific focus, one particular demand that they try to satisfy in a repetitive and rigid way.

What do entities want?
- It can be a particular desire or an addiction such as drinking, sex, or drugs.
- They can be focused on a particular emotion: pain, melancholy, suffering, punishment, guilt, violence, etc.
- They may just want to be looked after, taken care of and nurtured.
- Some just want to be left alone to hide inside, slumbering in a warm and snug environment.

Various other possibilities may occur, but one of the most common characteristics of entities is to be programmed in a narrow direction, as if to repeat the same message endlessly.

Case study 2.1 Thirty-six year old woman, unemployed.

Does it enjoy certain foods? –Sugar. Starch. Stuffing myself with bread and butter. He wants me to be fat. He doesn't want me to be attractive. He doesn't want me to be in a relationship. He wants me to be alone and masturbate, masturbate, masturbate! That's all he is interested in.

What happens to it when you masturbate? –It makes it full. It's as if he was making love to me. But he's never had enough. I have a very clear image of his face, now. He looks violent. Quite a violent sort of person, very mean. He looks a bit like my grandfather, but violent, and dark.

Entities

Case study 2.2 Forty-nine year old woman, shop manager.

What does it look like? –It's like something holding my jaw, something holding me back. It's dark and heavy. Like a metal clamp around my head, holding it. It's been there for a long time. When I used to get terrible headaches, it was the same vibration. It's like sadness and I don't want to feel it. It's as if it was holding me down. Like a very big weight, sitting on my shoulders. Like darkness.

What does it want? –I don't know. It's clinging onto me. It's very demanding. It wants something. It's cold. It wants warmth. It feels very cold and sticky. It's strange. I have felt that before, like an octopus type of thing, wobbly. It's sitting on my shoulder. It knows it's ugly but it wants to be loved. But I find it horrible. It has long arms, sort of clinging, like some of those sea animals. It feeds off me. It has a mouth and it attaches onto me.

Are there any foods that you eat that it enjoys? –Chocolate! [The client sounds quite surprised at the discovery.]

What happens to it when you eat chocolate? –Chocolate makes it feel warm. Isn't that strange?

No! –And it would like me to smoke, because that makes it feel warm. I remember now that each time I drink wine, I feel *exactly* the same pain in the jaw.

What else does it want? –It just wants to be kept warm and loved. It feels very much like something that is not part of myself. It's something that does not belong to me. It's just hanging on, as a big dark blob of negativity. The closest I can describe it is like a flat jelly-like grey dotted thing, with suction arms like an octopus.

What does it take out of you? –Warmth, or something that has to do with warmth. It seems to suck it out of my head, where the head is connected to the spine. It sticks to my back and my shoulders.

Case study 2.3 Forty-one year old man, plumber.

What does it look like? –I don't want to look at it.

You don't want to look at it or it doesn't want to be looked at? –It doesn't like you to look at it.

What does it want? –It's curled there. It just wants to be left alone in the dark. It's scared. It doesn't want to let the light in. It's like a black blob

The Facts (continued)

that drains my brain power. It just says: "Go away, leave me alone". But that's a bit much, frankly! It's MY body!
Could it be that there are some foods you eat that it enjoys? –Chocolate.
What happens to it when you eat chocolate? –It's satisfied. It sleeps. It doesn't feel malicious, just selfish. It doesn't want me to think too much so that I don't wake it up.

Case study 2.4 Thirty-two year old woman, engineer.

Are there foods that it enjoys? –No, it doesn't seem to be interested in food.
What does it gain out of being in your back? –Somehow it seems to get something out of my thoughts. When I work, my cerebral activity feeds it.
How does it react, when it is fed by your thoughts? –It makes it all happy and excited, like a dog when you put food on its plate.

2.2 Confusion

Entities nearly always create a certain degree of inner confusion, from which they seem to derive several benefits. The confusion manifests in two main ways. Firstly, entities tend to promote anything that goes against the client's clarity of mind. Clients often describe how their entity pushes them to eat too much or eat heavy junk food, have another drink of alcohol or abuse various other toxic substances, stay awake a bit longer instead of going to bed, spend time in noisy, disorganised environments, etc. Anything that decreases mental acuity and favours inner blurriness seems to be welcomed and favoured by entities. Similarly, many clients describe how their entity tends to become overanxious, agitated, threatened or threatening if they meditate or if they just try to remain silent and motionless.

Secondly, entities are expert at camouflage, making clients believe that they themselves want what the entity wants. This becomes obvious once the presence is identified: clients suddenly get the feeling that they are being tricked and manipulated by something foreign.

Case study 2.5 Twenty-five year old woman.

Is there something like a presence attached to it? –Yes, an awareness. Something watching me. Like a voice, passing judgments all the time,

commenting on everything. It tries to mess things up. It tries to embarrass me, making me feel stupid, telling me that I look awkward. But it's good too.
What's good about it? –It's very alert. It sees things as they are. It tries to make me think about things before doing them. But then it makes me think too much, and I never do anything. And it holds me back when I try to talk to people. I feel a kind of barrier. It makes me unclear when I speak. It's strong and it's weak. It's strong in trying to maintain itself, but it tries to make me weak.
So what's good for you, in all of that? –... [Confusion!]
Where would it go, if it was to leave you? –It would not go! It definitely feels foreign, but it's taken residence. And now it says it has become a part of me, whether I like it or not.
Do you really believe that? –Yes... No!

Case study 2.6 Thirty-eight year old man, cook.

What does it want, this dark cloud in your left hip? –It just makes me feel tired all the time. I just want to collapse and die. And be buried. The cloud would be happy if it could be part of the earth, of the soil. It doesn't want to be a person.
You mean you want to be buried, or the cloud wants to be buried? –I want to. I mean... No, it wants. Or maybe I want it because it wants it. It makes me want it...

Case study 2.7 Thirty-six year old woman, writer, who was exploring a witch-looking presence in the left iliac area.

What does it want? –She wants a place to be, a place to live.
What does it mean, for her, to live? –The excitement of churning and turning, and stirring things up. She likes strong emotions.
What happens to her when you have strong emotions? –She feels good. Alive. My emotions feed her, she lives on them.
Are there some foods that it enjoys? –She likes a lot of spicy hot food. That makes me sick. A lot of spiced meat and pickled stuff. And she likes it when I pig out, because it makes me heavy and sluggish.
What does she gain out of that? –She is happy because I'm not aware and she has full freedom. If I'm alert and know what I'm doing and what I want

The Facts (continued)

to do, then I don't get into emotional stuff. But if I'm sluggish and unaware, then it can happen to me really quickly, and she has control. Before I even know it, she can make me go into some emotional mess. It's not only a bloating of my belly that takes place if I eat too much, but also of my senses. I'm not clear or sharp. When I'm sharp, if I feel something wrong I can say I'm not interested in going into that. But when I'm sluggish, I'm not clear enough to see it coming and she takes over.

2.3 Entities are a poison of the will

The confusion and blurriness also affect decision making, and many clients describe how their entity is a major factor in procrastination. Entities often like to play on doubt and guilt. If they can push towards an irresponsible act, they gain benefit from the client's dismay.

Case study 2.8 Twenty-four year old man.

–It likes cigarettes. Because each time I want a cigarette, there is a conflict: should I have a cigarette, or should I not? It's like a wave of doubt, it makes me weak. It gives it more power over me.

Case study 2.9 Forty-seven year old man, second-hand car dealer with a heavy drinking problem. He arrived late for his appointment.

–When I was driving here tonight, Harry [the name the client had given to his entity] suddenly made me have an irresistible desire for an Indian meal, with spices. I knew I had to come here but it was stronger than me. I had to stop the car at an Indian restaurant. And while I was ordering, I heard myself asking for a bottle of wine, and I knew it was Harry. And then, in the middle of the meal, I realised that we couldn't do the session tonight because I had drunk. I was pissed off at myself but it's not the first time it has happened.

2.4 Entities don't like to be seen

It is common for entities to react negatively when they are discovered and observed. When they first look at their entity, many clients say that 'the thing' seems to be embarrassed, annoyed, threatened or even threatening. The entity has long used many subterfuges to remain unseen, and does its best to stay hidden. If entities foster confusion and blurriness, it is because it makes it easier for them to remain unnoticed.

Case study 2.10 Thirty-one year old woman, actress.

What does it want? –It's very afraid. It does not want to feel anything. Afraid of people. Afraid of you. It wants to live in the dark, unknown. It wants to be unconscious. It doesn't want to be seen. And it makes me not want to be seen.
That must be a problem, for an actress. –[Laughing:] Yes, it is! It wants to stay hidden and to keep me in a dark cellar, where it could control me. It's dirty and it doesn't want to be alone. It wants me to be like it, dirty. It likes it when I eat too much and drink alcohol and go to bed late.

Case study 2.11 Fifty-two year old man, public servant.

How does 'the thing' react when we look at it? –It hates it. It says if you don't leave it alone, it's going to attack you and to make you sick, and you will die.
Is that all? –No. It will pursue you after you're dead. It's fierce.

2.5 Secondary benefits and shared dispositions

During the exploration process it often becomes obvious that whatever problems and disorders the entity may create, the client also derives certain benefits from its presence. These benefits may vary greatly from one client to another, but certain themes come back with steady regularity.

The most frequent benefit clients describe is that the entity magnifies sensual pleasures of various kinds. As seen before, entities conspicuously crave sensual enjoyment and harsh emotions, getting a thrill out of them. This thrill reverberates in the client, and is perceived as an enhancement of the enjoyment or of the emotion. The entity's enjoyment and that of its host reinforce each other, so that eating chocolate or masturbating become acutely intense and addictive experiences. This makes certain clients quite ambivalent as far as their entity is concerned. They recognise it as a foreign thing, a parasite creating all sorts of inner problems they do not want. At the same time, however, they are attached to the enjoyment it favours and intensifies.

The Facts (continued)

Another benefit frequently related by clients is that the entity keeps them company. A number of clients have described how the entity first approached them during childhood, at a time when they felt lonely and yearned for companionship. The presence entered and filled an emotional vacuum. The negative aspects such as energy drain and parasitic emotions were realised only many years later.

While exploring entities, it is not unusual for female clients to report that the presence satisfies their need to take care of someone, for nurturing emotionally and physically (see Chapters 6 and 7). In such cases, a subpersonality seems to subconsciously react to the entity as if it were the woman's baby.

In most entity cases, it is not very difficult to find some kind of similarity between the entity's desires and the dispositions of the client. The entity and the client often share certain emotional inclinations, or an addiction, or some other feature. Whatever the pattern may be, when comparing the nature of the entity with the psychology of the client, it makes a lot of sense that the former ended up in the latter.

Case study 2.12 Forty-nine year old woman, travel agent.

–It bites my neck... in the middle of my neck. It's got lots of legs and they've got suckers.
What part of yourself could benefit from the presence of this thing?
–It needs me to feed off. That makes me useful for something.
Is it nice to feed it? –No, it's not nice. But it makes me feel that something wants me.

Case study 2.13 Fifty-three year old woman, music teacher.

–It lives in my belly and it's the boss. I'm not the boss. That's the place where the voices come from. I have no control over it. Things happen out of control... and then I think that it's God that is punishing me because I did something naughty.
What does it look like? –It's just like a big black lump. It has its own ways, it's the boss. It makes me EAT. It makes me voracious. It's a battle between it and me but I'm weak. It's stronger than I am. It says it wants to punish me.

Entities

> *What's the part of yourself that gains benefit from the presence of that thing?* –It's the boss. I don't want to be responsible. I want someone to be responsible for me.

Case study 2.14 Thirty-six year old woman, fashion designer.

What are you feeling? –Mm... I can see something but it's ridiculous. Like an old witch. A sort of old, old, ugly person, dressed in black. In the darkness, there [indicating her left iliac area]. But she'd like to get into my heart.

What does she want? –I think to take something away. She makes the area weak. She puts darkness. She lives inside somewhere. She makes me angry and I make her angry. She is pretty nasty. She likes to be miserable, grumpy.

What sort of things does she like? –I don't know... Probably some types of sex. Really, mm... I suppose we could say dirty sex. It amuses her. It's a sort of mixed feeling. Because I don't like her. But at the same time I don't feel wrong about dirty sex. So it sort of confuses me. She is a bit naughty.

A bit? –A lot!

What proportion of your sexual desire comes from her? –Probably 50%, 60%. I don't think she is 100%. She makes me want to see people... that I know I should not see. She makes me be very seductive and attractive to people I should not see. Like this young man. She has used him. Scared him. Frightened him. She plays tricks on a lot of men, actually. And she can also frighten them.

Do you sometimes let her play in your imagination? –Mm... Yeh! Of course. It can be very exciting. And when I have sex she participates. She changes her image. She can be incredibly beautiful. Another face. She makes me quite seductive.

What type of food does she like? –That part of me doesn't eat much! She likes alcohol and junk food. Perhaps she likes quite sophisticated foods too. Yes, I think she could be quite sophisticated.

How does she react to your children? –She makes me be sexual when I relate to them. My son tells me off.

Did she like it when you had your cancer of the uterus? –Yes, very much. It was like she was taking her revenge, because I was not doing what she wanted me to do. Actually, I think she plays tricks on me and makes me think I love her much more than I really do. I'm a bit confused...

The Facts (continued)

2.6 Voices

Certain entities give clients a sensation of voices in their head. Contrary to some cliches, this does not apply to all entities. According to clients' reports, only a very minor percentage of entities create such voices.

Case study 2.15 Forty-four year old woman, writer.

–I have a sense of confusion. It does not feel like my confusion. It's very distracting.

What does it want? –It wants me to do things for it. It wants me to do the things that it wants, and that it's not been able to do. This is a very old feeling: being acted on in some way, being taken in certain directions which are not what I want myself. And I feel a mixture of pity and of being bound.

What sort of things does it want? –It wants some kind of acclamation, success, praise. It's hungry for those things. It looks like a spectre, hooded, grey and miserable. I don't seem to be able to turn away from it. It was very ambitious itself, it wanted to be famous on a large scale. But it didn't have the ability, it didn't have the gift to do that. But it says I have the talent it didn't have and I'm not using it and it's wrong of me. This thing makes me feel driven. It's such a miserable, lonely thing. It's always been an outcast, always on the outside, but not because it wanted to be. Because it somehow couldn't manage with people or situations. I don't know where it came from. It's something that latched onto me just because it knew I could.

What sort of benefit do you gain from its presence? –Familiarity. I'm used to it. The dialogue with it is very familiar. And it's comforting to know that it's always there. It commiserates with me, as if we are fellow-sufferers.

Could it be that there are some foods you eat that it enjoys? –Very salty things. Hams and salty meats, and strongly flavoured things. It needs to be fed such a lot! It's never filled up. It always wants more. It's never ever peaceful.

–It's got me frantic about my writing to the point where I can't write any more. The urge for writing was extreme. It was kind of a driven thing. Now I feel too frightened to write at all. It's as if I realised I was doing it for the wrong reasons. When I think of writing now, my mind goes into a state of complete anxiety. The spectre goes on saying: "You have the talent, you must do it". That stops me even more. The spectre says: "Do it, do it." And then: "See, I knew you couldn't do it." Then it is satisfied.

2.7 Feeling of being watched

Case study 2.16 Nineteen year old woman, student.
–It's looking at me all the time, passing comments on what I do. When I start something, it tells me I have no chance of succeeding. If I succeed, it finds all sorts of reasons why it is not really a success. If I fail, it tells me it's normal because I am ugly and stupid. The presence never stops putting me down.
What does it gain out of it? –It makes it happy for some reason. It makes it feel full. And it's always looking at me, watching me. It makes me feel self-conscious because I know it is always keeping an eye on me whether I eat, talk to people or masturbate.

2.8 Many entities are stronger when the client is alone

In a number of cases I have heard clients comment that their entity has more influence on them when they are alone. Some entities even patiently wait for the client to be alone before making themselves felt.

Case study 2.17 Forty-one year old man.
–As soon as the last person has left and I'm alone in the house, I can feel it become manic. Even if I'm not thinking about it at all, it takes me by surprise. Just because I'm alone it gets all excited. It sends me voices: "Lie down, lie down!" It's like it pulls my hands to my genitals. The sexual desire is very much exacerbated.

2.9 Physical disorders

Case study 2.18 Forty-eight year old woman, health practitioner. A few hours after going to the hairdresser, she started having an excruciating migraine that took eight days to subside. Then she had sex and the migraine started again at the time of orgasm. Nothing, from homoeopathy to morphine, seemed to alleviate the pain.

The Facts (continued)

What does it look like? –It's like a big insect, or something like a crayfish, on the left part of my skull. It certainly seems to have legs that are dug in around my occiput and my eyebrows.
What is it doing there? –It's feeding. It's got a little tube going inside my head and it's feeding like a tick. But it's not sucking blood, it's sucking some sort of energy. It's feeding off my anger and it's getting bigger as it feeds.
What does it want? –It's there to confuse me. It stops me from thinking clearly. It came when I was in the salon, while the hairdresser was cutting my hair. The migraine comes out of the tick like a poison. It comes out of its belly. It seems to be injecting it under my skin. And then the pain is terrible.

This client's headaches stopped immediately after the clearing of that entity.

It would take several treatises to exhaust the topic of physical disorders and illnesses related to entities. I will just give a few simple facts, as I have observed them in my clients.

Not all entities create physical troubles or illnesses, and it would be unrealistic to try to incriminate them in all diseases. Nevertheless, while in the ISIS state of expanded perception, a number of patients get the absolute conviction that their illness or physical disorder is due to an entity, as in the example just given.

There does not always seem to be a logical reason why certain entities create diseases or physical disorders, while others never do.

In the case of a chronic illness, when an entity is identified and found responsible for the illness, the patient often reports that the entity had been present for a long time before the illness appeared. It first created a functional disorder, which gradually crystallised into the illness over the years. Clients themselves often comment that if the entity could have been cleared right at the beginning, the disease could probably have been avoided.

It has also been my finding that when an entity behind a physical illness is discovered and cleared early, the results can be spectacular. However, once an illness has evolved for some time, it seems to gain a momentum of its own. Then clearing the entity is, in itself, often insufficient to bring about healing in the client.

Entities

2.10 Latency

While in the ISIS state, many clients come to the realisation that their entity has been inside them for years, if not decades, in a latent state. The entity was dormant like a seed, waiting to blossom into a full entity.

> **Case study 2.19** Sixty-two year old man, retired engineer. He was operated on for cancer of the sigmoid colon one year before. When he came to consult me, the disease seemed to have started again, with heavy pain in the left iliac area. This man came from a strictly rational intellectual background, and it was a great surprise for him to discover such a 'thing' in his bowels.
>
> *What does 'the thing' look like?* –It's alive. It's moving. It's grinning at me. It's not a human face.
> *What does it look like?* –It's grinning, showing its teeth, standing away from me, not letting me touch it. It has no body, just a face and teeth.
> *Is it the first time you see that?* –Yes. It's the first time I let myself see it.
> *What's the connection with your pain?* –The face is causing the pain. It controls whether I have the pain or not. It's beating me.
> *Could it be that there are some foods that it enjoys?* –It enjoys tomatoes and... tomatoes, tomatoes!
> *What happens when you have tomatoes?* –It smiles because it gives it a chance to beat me. It was inside me long before I got sick, but the sickness allowed it to break out. It didn't create the cancer, but it benefited from it. It might never have been able to express itself without the cancer.[1]

2.11 Entities like earth lines

In many cases I have observed a connection between entities and noxious earth lines (also called earth-ray lines, or sometimes ley lines). When a toxic earth line is found crossing the bed of a client who has an entity, the body part in which the client describes the entity is often just on the earth line during the night. Moreover, people with nasty or violent entities are often people who live in a house full of toxic earth lines.

It virtually never happens that an entity comes back after having been cleared. In some rare cases, however, it may happen that in the weeks

---

[1] In anthroposophical medicine, tomatoes are strictly forbidden in cancer cases.

The Facts (continued)

or months following the clearing, another entity invades the same body part. Whenever this takes place, one should immediately suspect toxic earth lines in the client's house.

2.12 Entities are particularly tenacious

Entities do not go away unless a special clearing technique is applied. One can implement any form of psychotherapy on them, from psychoanalysis to regression — it is extremely unlikely that the entity will disappear. All these methods of psychological investigation may allow clients to understand why and how the entity came in, and help them live with it better. In my experience, however, they have proved conspicuously ineffective as far as the main point is concerned — getting rid of the entity. This finding not only applies to my own practice; it has been confirmed by a number of therapists linked to the Clairvision School.

Among hundreds of cases of clients presenting a similar 'syndrome' to the ones described in this chapter, I can recall only a handful whose entity apparently disappeared without a proper clearing. Even among these, almost every time there were doubts about whether the case was a genuine entity or some fancy created by the client's mind.

On the other hand, when a proper clearing is implemented, in nearly all cases the entity disappears immediately, and never comes back. This applies to all clients and all entities, but there are important restrictions on who is capable of performing the clearings. The clearing process appears to me as a delicate and potentially dangerous operation (both for the clearer and the client). It should therefore only be performed by people who have particular skills. This issue will be discussed further in Chapter 15.

Nevertheless, when a clearing is performed by a qualified clearer, the results are excellent. During the clearing itself, many clients describe how they can see the entity being lifted up out of their body. Even if they cannot see it, almost all get the clear feeling that 'something' has left their body. After the clearing, clients can no longer feel the presence of the entity, and most of the symptoms associated with it subside or even disappear. The voices, if there were any associated with the entity, stop. The mental blurriness gradually clears. So does the fatigue. Clients describe having more energy and enthusiasm. The cravings do not necessarily disappear, but lose some of their compelling character.

I am *not* suggesting that all the client's problems are solved by clearing an entity! Some of the symptoms directly related to the entity's presence will disappear. Other psychological troubles will remain after the clearing. Even though it may become easier to deal with them, clients still have to face their own issues.

2.13 Entities and subpersonalities

When considering the topic of entities, one important question usually comes to mind: are these entities not simply parts of the client's psyche? Are they not merely unresolved complexes, subpersonalities, or parts of the client's 'shadow'? Before going further, I will point out the main reasons why I consider this *not* to be the case.

Clearing entities represents only a fraction of my practice as a therapist. During the last ten years, I have applied a whole range of methods of psychotherapy and psychological exploration in my work with clients, in particular the ISIS technique. This has involved, among other things, extensive work on the exploration of subpersonalities and characters. My observations have led me to recognise a fundamental difference between a subpersonality and an entity: a subpersonality is self-generated. It is a part of the client's psyche that has gradually been formed and developed over the years. An entity, on the other hand, is something which has suddenly entered the client. In many cases, clients can actually re-experience the moment when 'the thing' came in. They can see how it came from outside, approached them and found its way inside.

A second important reason is that when an entity is discovered, it is perceived by clients in quite a different way from the way subpersonalities are perceived. In some phases of the ISIS process, clients undertake a thorough exploration of their subpersonalities. They learn to identify them systematically, and watch them in their daily life. This gives clients a certain familiarity with the theory and practice of subpersonalities. When dealing with entities, however, their subjective experience is different. In particular, an entity gives a much greater feeling of separateness than a subpersonality. When exploring an entity, clients often make comments such as "It does not feel like myself", "It feels foreign", "It is a parasite", "It has not always been inside", "It has taken residence, but it does not belong here". Moreover, subpersonalities are considerably more complicated than entities. An entity is usually an unsophisticated 'lump' of energy and consciousness with simple and predictable behavioural patterns. There are exceptions to this rule but they are relatively rare. Subpersonalities, on the other hand, are much more complex parts of the client's psyche, with multiple ramifications in their mental and emotional structure. Besides this, practitioners of the ISIS technique always seem to establish a clear distinction between the experience of entities and that of subpersonalities. One

The Facts (continued)

may argue that these are subjective experiences — but so is the psychological field in general.

In this process, an essential element is the clearing that concludes the work on an entity. During the clearing, which is described in Chapter 15, clients often *see* the entity being expelled from them, and taken into the light. Even if they do not see it, they often feel that 'the thing' is being lifted up and leaving them. Afterwards, they can no longer feel the presence and in a matter of days, a number of the related symptoms significantly decrease or even disappear. Thus, in my experience as a therapist, the work on entities has proved a remarkably quick and efficient way of improving the health and well being of a number of clients.

Last but not least, entities can be perceived clairvoyantly. If someone was to ask me why I 'believe' in entities, my answer would be simple: "I believe in them because I can see them!" Furthermore, during years of work at the Clairvision School, clients, students and instructors have reported hundreds of similar clairvoyant observations of entities.

CHAPTER 3

TAOIST PERSPECTIVES ON ENTITIES

For our study of entities to be fruitful, we need to introduce certain patterns of understanding. One point must be made very clear; I do not suggest that these patterns explain everything about entities, and even less that they 'prove' their existence. The purposes of the three coming chapters are quite different. Firstly, to examine how entities have been understood in other cultural contexts, such as traditional Chinese thought, and Hinduism. Secondly, to see how some of these views can be related to our case studies, and help discern patterns in the behaviour of entities.

3.1 Xie Qi, 'perverse energies'

Let us start with the concept of *Xie Qi*, or 'perverse energy', as developed in traditional Chinese medicine.[1] To avoid any confusion, let me state from the outset that entities and perverse energies are not the same thing. However, the concepts of *Qi* and *Xie Qi* are worth examining, for they will help clarify a number of points related to entities.

Qi

Since our present purpose is to penetrate ancient Chinese thought, I will speak of acupuncture as it was understood from a Taoist perspective, and not as modern commentators try to explain it 'scientifically'. From a traditional point of view, acupuncture is based on harmonising the circulations of what the Chinese called *Qi*. The literal meaning of the word *Qi* is 'breath'. However, the concept of *Qi* goes far beyond the process of respiration. In Chinese medicine, *Qi* refers to an energy which is not physical, but which sustains the vitality and functioning of the physical body. The concept of *Qi* is very similar to the *prāṇa* (life force) of the Hindus. Just as the Chinese word *Qi* means

---

[1]*Xie* is pronounced 'kseeay', *Qi* is pronounced 'chee'.

Taoist Perspectives on Entities

'breath', so does the Sanskrit word *prāṇa*. Just as the *Qi* is seen by the Chinese as the principle that keeps the physical body alive, so is *prāṇa* in the Indian tradition. Just as any disorder of the *Qi* can result in illness, so can any disorder of the *prāṇa*. Just as the *Qi* is permanently circulating throughout the body in the meridians, so is the *prāṇa* in the *nāḍīs*.

In order to understand the concept of *Qi*, two points must be emphasised. Firstly, the *Qi* is not a physical energy. Therefore, from a Taoist perspective, modern attempts to reduce acupuncture to some kind of reflex action of the nervous system do not make sense. The nerves are part of the physical body, made of flesh, bones, vessels, etc. The Chinese perceived the *Qi* as an energy of a more subtle nature than any of these physical structures. The *Qi* permeates all parts of the physical body, gives them life and makes them function—but it cannot be reduced to any of them.

The second essential point is that not all *Qi*, or 'energies', are regarded by the Chinese as being beneficial to health. On this topic, people are sometimes confused by the word 'energy'. They think: "If it's energy, it must be good for me. One can never have too much energy." This completely overlooks the qualitative side of the *Qi*. There is actually a great diversity of *Qi* energies, some favourable to the body, others toxic. An attitude that says "any *Qi* must be good for me" could be compared to saying "my physical body is made of physical substance, therefore any physical substance must be good for my body." Take just a little bit of the very efficient poison called cyanide, for instance, and it is

Xie Qi

the end of your physical body! Just as many physical substances are toxic if introduced into the body, so a great variety of *Qi* can be toxic too. *Xie Qi*, or 'perverse energies', is the name given to these toxic energies by the Chinese. In *Xie Qi*, *Qi* means energy or life force, while *Xie* means vitiated, corrupt, irregular, heretical. *Xie Qi* refers to unhealthy environmental influences which weaken the system and can create diseases.

Xie Qi, perverse energies, are not all lethal of course, just as only certain toxic physical substances are fatal. For a long time, many perverse energies may not even create disease, or even a functional disorder in the physical body. They create however, a functional disorder in the *Qi* of the body, a disharmony which upsets the balance of the *Qi* layer. According to the Chinese, the earlier one corrects such a disorder of the *Qi*, the more one secures a prevention of disease. For in most cases, before a disorder crystallises into a physical illness, it first exists as a disorder of the *Qi*. By recognising and harmonising imbalances of the *Qi*, one avoids many diseases. This emphasis on preventive medicine was reflected in the fact that the ancient Chinese paid wages to their doctors only as long as they were healthy, and stopped paying when they were sick—a concept which all health ministers would do well to ponder on!

In 'perverse energy', the adjective 'perverse' can be misleading, for it gives the idea of something intrinsically wrong, if not evil. Perverse energies are none of that. They are just in the wrong place when in a human body. A much better term would be 'inappropriate energies', or 'misplaced energies'. Let us illustrate this point with two examples. Suppose you are given only carbon dioxide to breathe, and no oxygen — you die. For you, the carbon dioxide has acted as a lethal (physical) 'perverse' factor. However, the very same carbon dioxide is fine for a plant to breathe. On the other hand, give only oxygen to the plant, instead of carbon dioxide, and the plant dies of suffocation. Thus what supports certain living systems is toxic for others.

Another example is that of bacteria in a compost heap. To alchemists, compost has always been a wonder and a subject for deep meditations — it is based on rotting, and yet it is the foundation of the fertility of Nature. Think of all the beauties of nature: grass, trees, flowers, fruits. All of these rest on the rotting that first takes place in the soil. The greater the rotting, the greater the fertility of the soil. And the more bacteria, the more rotting. Now suppose some of the compost heap's bacteria find their way into your digestive tract and make you sick. For you, these bacteria have suddenly become a perverse factor. Yet they are exactly the same as those which were so beautiful in the compost heap. There is nothing evil or bad about these bacteria, they are just in the wrong place when proliferating in your intestines. The very same applies to perverse energies. There is not necessarily anything intrinsically wrong with them. They are just in the wrong place when attached to your *Qi*.

Certain watery types of *Qi*, for instance, are perfectly normal in a marsh. However, if the same energy gets mingled with the *Qi* of your liver and remains embedded there, an uncomfortable feeling may result, possibly accompanied by digestive sluggishness, and even fatigue. You will not necessarily be 'sick', but you may well feel there is 'something wrong' in your liver. Many people actually have the feeling of something clogged or cloudy in their liver, like a blurriness they can't clearly define. They go from doctor to doctor, and none can ever find anything wrong with them, for the disorder is not (yet) on the physical level. It is just a disharmony of the *Qi* layer.

Perverse energies are not the same things as entities. However, they share certain similar characteristics. For instance, both perverse energies and entities are foreign elements that cannot be properly digested and integrated. Therefore they remain inside the human system as some kind of foreign, non-physical thing. We will come back to some other common points later on. As we will see, one of the major differences is that entities are far more difficult to dislodge than perverse energies.

3.2 The Po and the Hun

The main Chinese word for entity is *Kuei*.[1] In order to understand how the ancient Chinese viewed the nature and the origin of the *Kuei*, one must first understand how they conceived the human psyche. According to them, human beings do not have one soul, they have ten. In certain texts, the number is even greater — up to *Bai Shen*, 'one hundred souls', which can also be understood as an indefinite number of 'soul parts'. Most traditional texts, however, limit themselves to ten souls, or 'soul parts': seven *Po* and three *Hun*.

The *Po* are sometimes also called the seven emotions: anger, desire, fear, joy,[2] grief, love and hatred.[3] In other words, the *Po* refer to the part of the psyche in which conditional emotions arise. The *Hun*, on the other hand, refer to the more spiritual parts of human sensitivity and intellect. Thus the *Po* are said to be gross, opaque and heavy, while the *Hun* are said to be light, translucent and subtle.[4] For a human being to come to life, there must be a combination of *Po* and *Hun*. Incidentally, the *Po* arrive first. It is even said that it is only when the baby gives its first smile that the *Hun* have arrived. However, the development of the *Hun* is not completed before the age of thirty. Emotional upheavals correspond to a turmoil of the *Po* layer, which in turn affects the *Hun* negatively. Thus, in Chinese, a poetic way of expressing a great fear is: "my three *Hun* are fidgeting and

---

[1] *Kuei* is pronounced 'kway'.

[2] In Chinese Medicine, joy is classified among the potentially dangerous emotions, in the sense that it may create an inner turmoil responsible for various health hazards, such as a heart attack.

[3] Interestingly, in Chinese the same word *Po* is also used to indicate the dark fraction of the lunar disk.

[4] In Pinyin, *Po* is transcribed with a descending accent and *Hun* with an ascending one. The word *Po* is therefore pronounced with a descending intonation, and *Hun* with an ascending one.

my seven *Po* are totally confused."[1] The individual's life on earth continues as long as the *Po* and the *Hun* are kept together in the body.

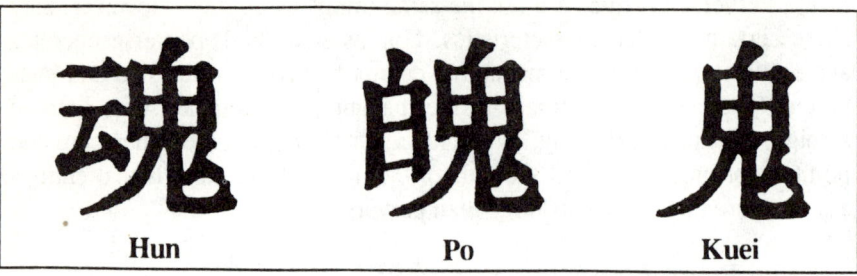

Hun   Po   Kuei

The doctrine of *Hun* and *Po* introduces an essential concept: that of the multiplicity of the psyche. In western languages, it is implicitly accepted that the psyche, or soul, is one. This is reflected in expressions such as 'my soul', 'a poor soul', 'a lazy soul', 'a lost soul'. There is obviously one soul per person — not more. In order to take into account the great diversity of people's reactions and patterns of behaviour, one sometimes discerns several subpersonalities, or characters, within the same individual. However, these characters are implicitly understood as being different aspects of the same psyche.

The Chinese, on the other hand, not only had a system in which subpersonalities were taken into account; they went one step further and considered that there was such a lack of unity between these subpersonalities that it was impossible to group all of them under one single soul. They divided the human psyche into seven *Po* and three *Hun*, thereby acknowledging that each of these parts had an independent existence. The *Po* and the *Hun* are of radically different natures. They come from different places and, as we will see, have different destinations after death. They are not different parts of one soul, but separate 'things' which have come from different spaces. They are held together only as long as the individual is alive, and they separate again at death.

Moreover, human existence is described by the Chinese as a permanent antagonism between the *Hun* and the *Po*, in which the *Po* are constantly trying to suppress the *Hun*. The proportions are clearly stated (seven against three), which is an interesting way of quantifying human nature. However, it would be far too simplistic to try to limit this antagonism between *Hun* and *Po* to a fight between 'good' and 'bad' psychological tendencies. For it is not only hatred and anger which are on the *Po* side,

---

[1]Henri Maspero, *Le Taoisme et les Religions Chinoises*, Gallimard, Paris 1971, p. 206

but also (conditional) joy and love. This indicates that the *Hun-Po* perspective is not just a moral one, but is based on a deeper understanding of the variety of parts constituting a human being.[1] This is not to say that there is nothing permanent or eternal in a human being as seen by the ancient Chinese. In their system, the *Shen* can be regarded as the equivalent of the Higher Self, or Spirit. However, the Chinese had a clear perception of the separation between Spirit and soul — or rather Spirit and souls. They saw their souls as parts of themselves just as we see our organs: important, indeed, but separate, transient, and to a certain extent replaceable.

3.3 *Kuei, or entities*

As long as an individual is alive, the *Hun* and *Po* are held together, and their interactions result in the varied modalities of psychological life. However, at death, the *Po* and the *Hun* are said to separate, in a way that parallels the astral shattering described in the next chapter. In the Chinese model, after death, due to their descending polarity, the *Po* remain bound to the earth. Whereas the *Hun*, due to their upward moving tendency, ascend to the spiritual worlds, following the *Shen* (equivalent to the Higher Ego or Self). After death, once the separation of the different parts constituting a human being has taken place, these parts are no longer called *Po* and *Hun*, but *Kuei* and *Shen*.

---

[1] In the pattern of subtle bodies which I use in *Regression, Past Life Therapy for Here and Now Freedom* and in this book, the *Po* emotions correspond to the mechanical movements in the astral body, i.e. the conditional emotions of the reacting mind (*manas*, in Sanskrit). The *Hun* correspond to the 'feelings' and to the transformed astral body (the Spirit Self of Steiner).

In Steiner's writings, the word 'soul' is strictly equivalent to astral body. In looser language, however, the *Po* could be translated as 'mortal soul', and the *Hun* as 'immortal soul'. A quotation from Goethe's *Faust* illustrates perfectly the concept of *Po* and *Hun*:

> Two souls, alas! dwell in my breast,
> Each wants to repel the other.
> The one clings to the world
> With tenacious clutches and lustful desire;
> The other rises mightily from the dust
> Into lofty ancestral spaces.
> *Faust*, I.ii (verses 1112-1117).

Similarly, after death the *Po* cling to the earth, and the *Hun* ascend into higher spiritual regions.

Entities

Shen

Shen denotes the Spirit, the transcendental part of a human being.[1] The word can also mean 'God', the 'Supreme Spirit', all the 'celestial spirits', and also 'miraculous' or 'supernatural'. However, due to slack habits of language, *Shen* can also mean 'soul', in the vaguest sense. For instance, when talking about the ten souls, or the one hundred souls (*Bai Shen*), it is the word *Shen* that is used.[2]

After death, it is said that all or parts of the *Hun* follow the *Shen* in its ascension.

The rest of the souls, i.e. the *Po* and possibly certain parts of the *Hun*, remain on the physical plane and become the *Kuei*.[3] Thus the *Kuei*, or fragments of the psyche of the dead, remain in the sphere of human existence, or very close to it, while the *Shen* and the *Hun* depart for spiritual worlds which are completely beyond the reach of normal human beings. The ancient Chinese spoke of the *Kuei* as 'wandering souls' or remnants of the dead—unsatisfied greedy spirits. The *Kuei* need to be fed and pacified with offerings, otherwise they start harassing the living. Chinese literature abounds in stories of *Kuei*, of which the following is an example.

Once a horseman of the imperial guard was hunting in the countryside when he passed near a well, where an old man was drawing water. The horse suddenly went out of control and started galloping frantically. Before the guard could do anything, the horse hit the old man and flung him into the well. A few seconds later, the horseman regained control of his animal, and raced away.

The story did not stop there. The following night, the old man appeared as a spectre in the guard's house, and started breaking all the plates, slashing the paper windows, insulting and terrorising the whole family. Everyone immediately understood that it was a *Kuei*. The guard tried to apologise: "It was not me. It was my horse who threw you into the well."

---

[1] In the Pinyin transcription of the Chinese language, the word *Shen* is written with an ascending accent, just as *Hun*.

[2] Similarly, in western languages, one often uses the word 'soul' casually, when in reality it is the Higher Self, or immortal Spirit, which is meant.

[3] *Kuei* is transcribed *Gui* in Pinyin. *Kuei* also denotes the twenty-third of the twenty-eight Chinese lunar mansions, including eta, theta, gamma and delta of Cancer. See *Dictionnaire Français de la Langue Chinoise*, Institut Ricci, Paris 1976, ideogram no. 2832.

The *Kuei* answered: "Shame on your ancestors, you filthy scoundrel! Maybe it was your horse who threw me into the water, but you just ran away instead of helping me to get out!" And the poltergeist-*Kuei* kept on breaking everything it could find. The whole family were prostrating themselves, promising they would give regular offerings to the *Kuei* if he would forgive them. "That is not enough," said the *Kuei*, "I want you to make a tablet with my name, and conduct exactly the same rituals for me as you do for your ancestors." They agreed. The *Kuei* gave his name, and a tablet was made accordingly. Every day, the family performed a ritual and fed the *Kuei* with their offerings. And the *Kuei* kept quiet.

In the years that followed, the guard cautiously avoided the area of the well. One day, however, while escorting the emperor, he had to visit the same place again. To his great amazement, the same old man was near the same well. The guard was panic-stricken, but his duty forbade him to flee. The old man immediately recognised him, rushed at him and threw him off his horse. The guard was so afraid that he did not dare defend himself. The old man beat him with a stick. "There you are, ruthless bandit!" said the old man. "At last I've found you!" All the other horsemen were laughing at their comrade, who was a very big man, and yet was unable to fight back while the decrepit old man was calling him names and beating him up.

"But, but...", protested the guard, "didn't my family and I feed you offerings daily? Didn't we make a tablet with your name?"

"Offerings? What offerings?" said the old man. "Are you trying to insult me even more, by pretending I am a *Kuei*?"

Suddenly, the old man understood what had happened. He became quiet again, and let the guard go. The day after, he went to visit the guard's family. He had no difficulty finding the place, for the whole town was laughing at the horseman's misadventure. All the family members were terrified when they saw the old man coming. However, he reassured them: "Fear not. I am not a *Kuei*! Show me the tablet." After taking a glance, the old man exclaimed: "This is not my name!"

The whole incident was due to a malevolent *Kuei*. The *Kuei* had frightened the horse in the first place, and then had taken the appearance of the old man in order to play havoc in the house of the guard's family. By so doing, the *Kuei* had found a home for itself, and had been fed daily offerings for years.

The old man, who was a bit of a magician in his free time, promptly performed an incantation to get rid of the *Kuei*. Immediately, a sinister roar could be heard, half laughter, half agony. And the *Kuei* never came back.[1]

In ancient China, the *Kuei* were not only part of the literature or folklore, but also part of daily life. In all trades and areas of life, it was customary to take the *Kuei* into account and perform rituals to protect oneself against their pernicious action. As stated before, the art of medicine, for instance, recognised *Kuei* as one of the factors which can create disease. Thus, among the 361 points of the 14 main meridians of acupuncture, 17 have the word *Kuei* as part of their main or secondary names. Governments also took great care of spirits of all kinds, and systematically addressed them with pacifying rituals. These were considered indispensable to secure peace and prosperity for the kingdom.

What happens to the *Kuei* in the long run? According to the ancient Chinese, the *Kuei* are not immortal. Their existence is limited, and lasts only until their vitality is exhausted. Afterwards, they dissolve. What belongs to the earth returns to the earth, is composted, and reintegrated in the universal cycle of energies.

3.4 Subtle bodies in a nutshell

To allow a productive discussion on the topic of entities, we must briefly define a few terms.

The *Qi*, or life force, corresponds to what western esotericists call the etheric. The **etheric body** is therefore the envelope made of life force. It is the equivalent of the 'envelope-made-of-*prāṇa*', or *prāṇa-maya-kośa*, of the Indian tradition. Just as, according to the Chinese, the *Qi* permeates the whole physical body, so does the etheric body. One could take the example of a sponge (the physical body) permeated by water (the etheric body). However, the etheric body also extends slightly beyond the limits of the physical body. In Tibetan medicine, one sometimes does acupuncture on certain points situated beyond the limits of the skin, which is indeed painless!

The souls which the Chinese call *Po* and *Hun* correspond to what western esotericists call the astral body.[2] In short, the **astral body can be defined as the layer in which thoughts and emotions take place.** We

---

[1] This story contains a moral that is still valid for 21st-century entity clearers — one should never take an entity's identity for granted.

[2] Strictly speaking, the *Po* correspond to the astral body, and the *Hun* to what I call the transformed astral body.

could call it the vehicle of our mental consciousness, including our emotional life.

The immortal part of human beings, their Higher Self, is what we will call the Ego, or Higher Ego. Here, Ego is written with a capital E, to distinguish it from the little ego.[1]

This introduces a simple model discerning four parts in human beings: a physical body, an etheric body, an astral body, and an Ego. For our present purpose, we can consider the words Ego, Higher Ego, Self, Higher Self and Spirit as synonymous.

As a preliminary to the discussions in the following chapters, here is a diagram indicating the nature of perverse energies and of entities in terms of subtle bodies.

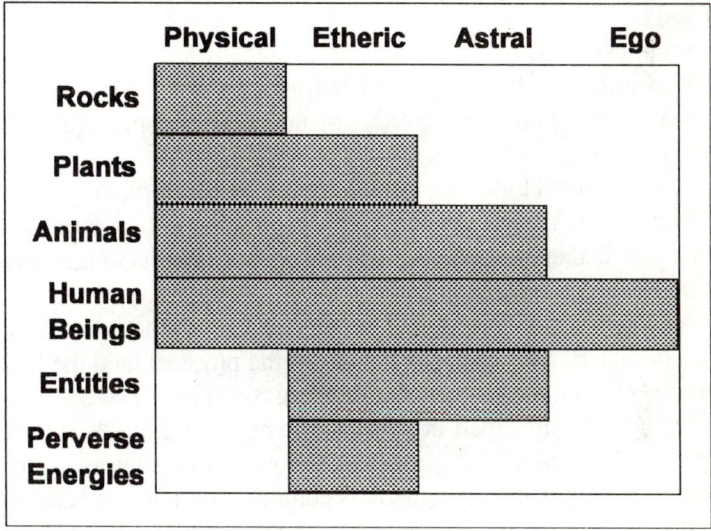

---

[1]In this model, the little ego, made of illusory conditioning and childish reactions, corresponds to the astral body.

CHAPTER 4

FRAGMENTS

4.1 Physical body–etheric body and astral body–Ego

In the last chapter, we briefly described human beings as fourfold, comprised of:
- a physical body;
- an etheric body, or layer of life force;
- an astral body, or layer of mental consciousness (thoughts and emotions);
- an Ego, (or Higher Ego, Self, Higher Self, or Spirit).

In practice, these four parts can be grouped into two pairs: on one hand, the physical and etheric bodies, and on the other hand the astral body and Ego.

As long as you are alive, the physical and etheric bodies remain closely associated. The etheric is the life of the physical, and the latter cannot survive without the former. On rare occasions, for instance when your arm or leg goes numb, a part of your etheric body temporarily moves out of the physical. The painful tingling that follows when starting to move the limb again, indicates that the etheric is coming back into the corresponding physical part. Such a separation is of course partial and transient. More complete separations are not totally impossible, but extremely rare.

There is, however, an exceptional situation in which a fairly high degree of separation of the etheric and physical bodies is said to take place. In certain initiations, as part of the esoteric training of ancient mystery schools, candidates were placed in a death-like trance for a certain time (classically three days), during which they were taken travelling into spiritual worlds. Once the three day journey was complete, the candidates were called back into their body. Because of what they had seen, they were called initiates.

However, apart from extraordinary circumstances of this kind, the physical and the etheric bodies never separate during life. It is only after physical death (due to the final departure of the astral body and the Ego) that the etheric body starts to dissolve in the etheric world, thereby abandoning the physical body, which starts to rot. So that, after death, the destiny of these two envelopes is quite similar; each of them dissolves and is reabsorbed by its environment. The physical body goes back to the physical world, while the etheric body goes back to the etheric world.

Just as the physical and the etheric bodies never separate under normal circumstances, so the astral body and the Ego are intricately entangled. From an experiential point of view, this corresponds to the fact that most people are unable to discern their Higher Self from their thoughts. The Ego (or Higher Self), is the layer of self awareness. The astral body is the layer of mental consciousness, that is thoughts and reactional emotions — and the astral body is entangled like a spider web around the Ego. Therefore, when people close their eyes and try to become quiet, they cannot contact their Self; they are unable to perceive anything but thoughts. The constant movements inside their astral body translate into permanent mental activity, and act as a screen which blocks off the Self.

In terms of subtle bodies, spiritual development can therefore be seen as a process aimed at separating the astral body from the Ego, by which one can become Self aware. In other words, **to find the Self, one must disentangle it from the astral body**. Or, to take an example often quoted in the Indian tradition, as long as Self realisation has not been achieved, Self and astral body remain mingled like milk and water in a glass.

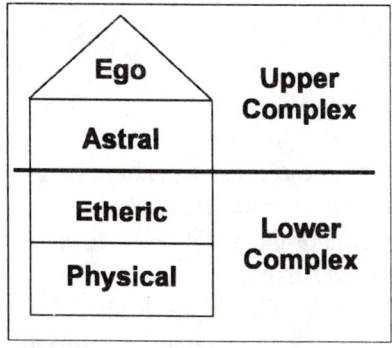

As far as the usual circumstances of life are concerned, it is possible to arrange the four vehicles into two groups: a 'lower complex', consisting of the physical and etheric bodies, and an 'upper complex' consisting of the astral body and the Ego.

4.2 Sleep, death and the astral body

To understand some of the key mechanisms related to entities, it is essential to get a clear picture of what happens to the astral body during sleep and after death. The most obvious fact about sleep is that it is accompanied by a loss of consciousness. Consciousness corresponds to the two 'upper' vehicles: the astral body (mental consciousness) and the Ego (self awareness). Therefore, in terms of subtle bodies, what happens during sleep is that the astral body and Ego lose interest in the physical and etheric bodies, and direct their activities toward different spheres. In reality, consciousness is not lost during sleep, just directed somewhere else.

During the waking state, the upper complex (astral body + Ego) is impacted in the lower complex (physical + etheric bodies) and cognises the physical world through it. During sleep, the upper complex becomes detached from the lower one, and directs its activity toward the astral worlds.

It should be clear that the multi-faceted reality of sleep and dream can only be roughly summarised by such simple diagrams. Several other mechanisms take place simultaneously. There can be various degrees of

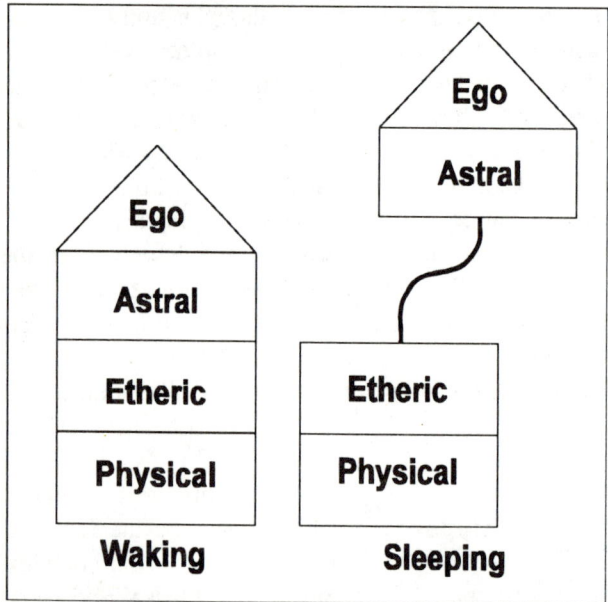

dissociation between the upper and the lower complexes, depending on the depth of sleep and the level of development of the individual. Separations may also take place between various sub-layers of the astral body. In short, a map cannot fully describe what a country looks like.

In terms of subtle bodies, the mechanisms related to death are an extension of those taking place during sleep. While sleeping, a transient separation of the upper complex (AB + Ego) and of the lower one (PB + EB) takes place. At death, a final dissociation occurs. The astral body and the Ego depart, abandoning the physical and etheric bodies. As in sleep, consciousness (which, in our system, corresponds to the astral body and the Ego) is not lost, it is just somewhere else. Instead of being connected with the physical world, it starts journeying in the astral worlds, and then in the worlds of the Spirit.

4.3 Functions and corresponding structures in the subtle bodies

Shortly after death, before the departure into the astral worlds, an essential process takes place: the shattering of the astral body. The astral body literally falls into pieces just as, in the Chinese model, the *Po* and the *Hun* separate. To understand this process, which plays a key role in the genesis of many entities, let us spend more time examining the structure of the astral body.

The concept of subtle bodies invites us to reconsider the way we look at various functions such as thinking, feeling and the emotional reactions we display from morning to night. From a conventional point of view, there is nothing more abstract and intangible than a thought. From the point of view of subtle bodies, however, a thought is a tangible form made of astral matter. It can be seen and even felt, 'palpated'—provided the proper non-physical organ of perception, the third eye, has been built. From this perspective, the fact that most people regard their thoughts as abstract and insubstantial only indicates their inability to perceive beyond the physical sphere.

Subtle bodies lead us to consider everything in terms of structure and matter—not only physical matter, but also subtle matter, including etheric and astral substances, and a whole range of even more refined non-physical substances. This focus on matter makes the system of subtle bodies the foundation stone of inner alchemy, for alchemy is fundamentally the art of raising the level of vibration of matter. Inner alchemy is a form of self-transformation work through which you become aware of the subtle frequencies of matter behind all functions, whether physiological, psychological or spiritual.

Thus in terms of subtle bodies, life is not an abstract principle, but the intrinsic quality of etheric matter. Life is the function and etheric matter is the structure. Similarly, on a higher octave, the whole range of psychological functions can be regarded as forms or waves in the astral body. What are we usually aware of, when experiencing an emotion? We perceive the emotion itself, that is the anger, frustration, dismay, etc., and we perceive certain physical modifications that accompany the emotion, such as muscular tension and quicker heartbeat. These physiological responses are obviously the consequences of the emotion.

Entities

What people usually do not realise is that the emotion itself is a consequence. The emotions people feel are not 'original emotions', meaning astral waves in their astral body, but the physical reflection of these waves. The emotion first starts in the astral body, and is then reverberated in the nervous system of the physical body. When people experience emotions, what they feel is the physical reverberation, not the original astral wave. This model, which altogether is quite similar to Plato's myth of the cave, gives clues as to why most people are unable to deal with their emotions: they simply never perceive them! They live in a world of effects and consequences, if not shadows, without any grasp of the real causes of their inner movements. As long as one deals with consequences and not sources, no real transformation is possible.

This reverberation process should always be remembered when trying to understand the astral body. The astral body is the layer of thoughts and emotions — the 'real' thoughts and emotions, not those people usually perceive. What they perceive are only reverberated thoughts and emotions, meaning the reflection of astral forms in their nervous system. So, if the astral body is the layer of mental consciousness, what people experience during their usual waking state is not mental consciousness but **'physical mental consciousness'**, meaning a physical reflection of the astral mental consciousness.

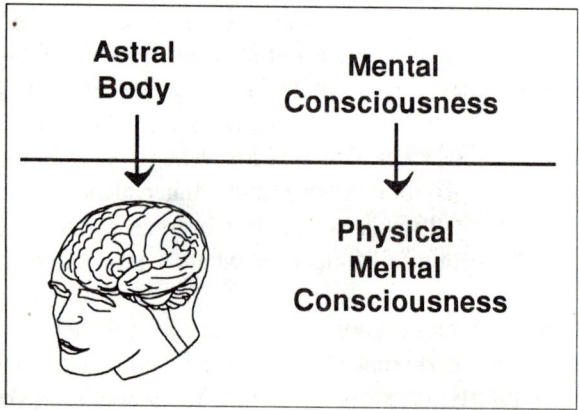

It is important to ponder upon this reverberation process, because a few major distortions take place in the reflection. For instance, a very obvious distortion occurs in that the astral body is an immense storage of latent emotions and memories, while the physical mental consciousness only receives a few thoughts and emotions at a time. In other words, the physical mental consciousness reflects only minute parts of what is contained in the astral body.

Fragments

Seen from the physical mental consciousness, the astral body is like a kaleidoscope presenting ever-changing patterns of thoughts and emotions, hence the ever-changing moods and thoughts of most human beings. From the perspective of subtle bodies, it is easy to understand why it is possible to suddenly feel very happy or very unhappy, without any external reason. The kaleidoscope effect is such that the physical mental consciousness is often redirected from one part of the astral body to another, and therefore reflects completely different emotions.

Some people are actually quite good at manipulating this effect. When they are disturbed by a certain emotion or desire, they just switch off. They turn the kaleidoscope and redirect their physical mental consciousness toward another part of their astral body, and forget about the problem. Actually the vast majority of the people who claim that they can 'transmute' their desires and emotions do nothing other than switch off. Instead of looking at them, they simply disconnect from them, redirecting the physical mental consciousness toward another less confronting part of their astral body. In reality, this has nothing to do with transmutation, it is only suppression. The desire or the emotion remains unchanged in some dark corner of the astral body and if anything, it is rather reinforced by being suppressed.

As long as one is alive, conscious existence takes place mainly in the physical sphere. This means that one is only aware of the physical mental consciousness, and that many hidden aspects of the astral body can only manifest through dreams or subconscious acts. At death, however, the physical nervous system stops functioning. The facade of physical mental consciousness therefore disintegrates, and conscious existence is projected into the astral sphere. Having lost their two lower vehicles (physical and etheric bodies), the dead go through the astral worlds with their astral body and their Ego. Now they can no longer cheat, they have to face the astral body the way it is, and in its entirety.

In practice, this means that all kinds of repressed desires and latent emotions rush into the consciousness of the dead. That is when the people who thought they had transmuted their emotions come to the bitter realisation that in reality they had never dealt with them, and that they even reinforced them by suppressing them. Latent emotions and desires suddenly become conspicuous. They appear in front of the dead as tempestuous and harassing forms. The experience is that of huge 'itches', burning desires for things one cannot get any more. Many esoteric schools have seen this phase of purification of the astral body as the 'purgatory' spoken of in religious writings. Note that with this approach, we are very far from the cliches of eternal hells full of little demons punishing the dead! Firstly, the purgatory phase is transient, and as soon as the cleansing process has been achieved, the dead continue their journey into the delightful worlds of the Spirit. Secondly, if the dead are harassed, it is by nothing other than their own desires and latent emotions. The process is quite mechanical, and devoid of moral connotations. Forces (emotions and desires) have been repressed and compressed in the astral body. As soon as the shell of the physical mental consciousness disintegrates with death, the compressed forces burst out like a jack-in-the-box.

Another important point is that the purgatory phase does not have to be an ordeal. It will be painful only if the astral body has not been worked on during life, and if it remained crammed with burning latent desires and emotions. But if a catharsis has been achieved whilst alive, if a systematic self-transformation work including an exploration of emotions has been conducted successfully, then the purgatory time will be quick and non traumatic. Furthermore, past a certain level of transformation, there is simply no more purgatory phase after death. Astral hang-ups have already been worked out, and so the transition into the Spirit worlds is smooth and unhampered.

In Sanskrit, the word for the latent tendencies of the astral body is *samskara*.[1]

4.4 On the astral level you are not a person, you are a mob

Another major distortion arises from the reflection of the astral body's mental consciousness into the physical mental consciousness — the latter induces a false sense of unity. Most people tend to think of themselves as one person, with their likes and dislikes, their desires and fears, and their emotions of various kinds. However, this impression of being one person does not rest on a perception of their true Self, of which they are usually completely unaware—for finding the Self is a long endeavour. For most people, the perception of themselves is in reality a perception of their astral body, or rather their astral body's reflection in their physical mental consciousness.

This is where a major misunderstanding occurs. If they were to see their astral body for what it really is, they would see not one person but many persons, which we could call subpersonalities or characters. The word subpersonality can be misleading, for it gives the idea of one single

---

[1] A systematic analysis of the mechanisms of *samskaras* and ways of dealing with them is the central theme of *Regression, Past-life Therapy for Here and Now Freedom*, by the same author.

personality that branches into various subdivisions. In reality, these subpersonalities are more like a crowd of characters which have very little, if anything, to do with each other. They are not like the different provinces that constitute a state, but more like birds of different species artificially kept together in a cage, competing and fighting with each other all the time.

Thus one character may be a lover of beauty and the arts and become ecstatic in a museum or with a picturesque landscape. Another character is power-hungry and pushes you to stay late at work and use any possible opportunity to achieve your ambitions. Another character would like to be a hermit and yearns for 'spiritual' seclusion in a monastery. But another character loves sex passionately, and could never make it in a monastery without being completely repressed. Each character has its own longings and dislikes, and aims at taking over your life.

Seen from a structural point of view –from the point of view of subtle bodies– this multitude of characters corresponds to a mosaic of astral parts. The astral body is made of 'bits'; it has no real unity. In ancient Greek, the meaning of the word character is that of a stamp, a mark that is engraved, impressed. Each character is imprinted in a particular bit of the astral body and there is hardly any communication, and even less cement, between the different bits.

For instance, if you decide to learn accounting and Japanese to reinforce the power-hungry character inside you, the data you memorise is imprinted in your astral body in a part related to that character. If you learn to play the oboe and spend your holidays in Florence, the life experience related to visiting all the museums, and the musical package, are engraved in a part related to the lover-of-beauty bit in your astral body, and so on. You may believe that it is the same person who learns accounting and who visits Italy's museums, but that is sheer illusion. In reality they are two different persons, cohabiting in your astral environment. It is only the fact that these two persons are reflected into the same physical mental consciousness that creates the illusion of being one single person. **On the astral level you are not a person, you are a mob.**

Some people manage to become quite successful in life by developing one particular character at the expense of all the others. One of their characters imposes its own desires and starts ruling as a dictator, and their whole life takes the direction imposed by it. For instance the power-hungry character takes over, and then no more holidays in the museums of Europe or plans for retreats in monasteries—only work. The more success the character meets with, the more it can reinforce and consolidate its position. Seen from outside, such people appear quite centred and 'mono-focussed', knowing exactly what they want and using all their inner resources to get it. However, in terms of subtle bodies, this does not mean that a unification has been achieved. The astral body is still a mosaic of bits. What has happened is just that one particular astral bit has grown out of proportion to all the others. The other bits are not harmonised with the dictator bit, they are just repressed and starved.

If there is one ruler that can harmonise and unify the mob of characters, it is the Ego (the Higher Ego, or Self, or Spirit). The more the Ego shines like a Sun at the centre of gravity of the astral body, the more the different characters start orbiting around it. Instead of working only to satisfy their own selfish desires, the characters start manifesting the purposes and the light of the Spirit. Instead of plotting for the success of their own ambitions, they start accomplishing the works of the Higher Self. This process also corresponds to structural changes in the astral body. The unveiling of the Self begins a unification process. A new astral body slowly develops. In this new, or transformed, astral body, the different parts are penetrated by the light of the Self. Therefore they are not only united around the Self, but also cemented to it.

These reflections on subtle bodies lead us to discern two factors of unity in a human being. One is fake, nothing more than an appearance: it is the illusion of being 'one person' because the multifaceted astral body is reflected in one physical mental consciousness. The other is the real unity that gradually develops with the realisation of the Self, and the pervasion of its light in all parts of the astral body.

In hermetic terms, the Sun stands for the Ego. The metal associated with the Sun is gold. The alchemical work of turning base metals into gold can be regarded as the penetration and unification of the astral body (and later on of all other vehicles) by the light of the Ego.

4.5 *The shattering of the astral body at death*

What happens at death? A final separation of the upper complex (astral body and Ego) from the lower one (physical and etheric bodies). The etheric body starts dissolving in the universal ether. Similarly, the physical body starts rotting.

The nervous system ceases operating and therefore so does the physical mental consciousness. And with the end of the physical mental consciousness comes the end of the illusion of being one single person. The dead suddenly realise their real astral nature—a mob of characters. The illusory shell of the mental physical consciousness is no longer there to artificially hold the astral bits together. The only cement that is left to hold the astral parts together is the Ego. Consequently, the only astral bits that will be able to stick together are those which, during life, have been penetrated by the light of the Ego. For most people, that is not much, because they have forgotten to search for their Self. Throughout their life, their Self has remained like a sleeping princess in the background of their personality—or rather of their mob of personalities. The Self has not been invited to take part in the life of the characters; the alchemical marriage of the astral body and the Self has not taken place. Only a few minute astral parts have been penetrated by the Ego. Consequently, the death knell tolls: most of the astral body falls into pieces. The mosaic of astral bits crumbles into astral dust and fragments that drift away in the astral space.

From an experiential point of view, this shattering of the astral body after death is quite a dramatic experience. Here you are, floating in the astral space, gradually being stripped of your astral substance. You see the part of yourself that could speak Japanese drop and drift away in one direction. Then you see the bit that liked to play the oboe leave you and drift elsewhere. And then the part that passionately liked sex falls out and departs in another direction. All these are like limbs of your astral body that drop off and start floating in the space. Apart from these main fragments, a significant fraction of your astral body just crumbles into dust, which spreads into the universal astral space.

Why do certain bits crumble, while others remain more or less intact, floating in the space as astral fragments? This has to do with how constructed and crystallised the corresponding characters were, that is, how deeply they were imprinted into the astral substance. Thus if one day you thought about learning the piano, and then bought one and casually tried to play for a few weeks before giving it up, the corresponding imprint in your subtle bodies is faint. After death, with the shattering of the astral body, the 'pianist bit' just dissolves, turning back into undifferentiated astral dust. On the other hand, if you have longed to own a piano for years, worked hard to be able to buy one and then passionately practised it, or if you started playing at the age of five and toiled on the instrument for hours every day, then the situation is quite different. Your dedication and your intensity have created a deep imprint in the astral substance. A structured, coherent and crystallised astral fragment has been generated. After death, when the fragment is separated from you, it will not fall into astral dust. It will remain as a pianist astral fragment floating in the space.

From what has just been said, we can understand that certain emotions and patterns of behaviour will tend to create more coherent and persistent fragments than others. The main criteria for crystallisation are repetition and intensity. Everything related to basic vital functions such as sex and nutrition, or addictions of all kinds, creates repeated and intense activation of certain parts of the astral body. These deep astral imprints tend to result in solid coherent fragments after death.

Suppose you were an alcoholic or a heroin addict, for instance. After your death, the part of your astral body that was addicted to the drug could well escape the general astral crumbling. The more intense the addiction, the more solid and persistent the astral fragment released. Or if you were a person that couldn't live without sex (and this applies to quite a few people nowadays), it is not because you die that the corresponding longing will be extinguished in your astral body. The astral fragment or fragments related to your sexuality will fall off from your structure, but will continue to long for sex while drifting in the space.

As discussed before, if during life there has been a strong desire that was repressed and buried deep inside the astral body, at the time of death it will reappear like a jack-in-the-box. Consequently, it is not only debauched people, alcoholics and drug addicts who release fragments with the most violent vital desires and passions, but also many so-called 'good' people whose life was based on repressing all their desires. A wonderful thing about death is that it does not allow any cheating. When dying, individuals are stripped of their facade, and their future destiny depends on what they really are, not on any appearance they may have built up during their life.

Of course, the astral fragments released after death are not all related to sex, food or addiction. Any strong mental or emotional disposition can create a fragment. Still, it is astral intensity that crystallises the astral bits. Take a good look at the 'astral intensities' (emotions, desires, etc.) of most of the people around you, and you will easily find out what the corresponding fragments will want after death.

In conclusion, if we look at the global destiny of the astral body after death, we can divide it into three parts:

–a small fraction of the astral body remains attached to the Ego, which departs for the journey into the intermediary worlds;

–a big part crumbles into undifferentiated astral dust;

–various bits break off and drift away in the space as fragments.

4.6 A short digression about reincarnation

The shattering of the astral body after death is quite consistent with the fact that under normal circumstances, the vast majority of people are unable to remember their past lives. When dying, most of the memories are lost in fragments or dust. What will reincarnate is the Ego, with the few

Fragments

shreds of astral body that remained attached to it. Before coming back into a new physical body, the Self will gather some fresh astral matter around it, to create a new astral body.

A memory is kept in the Ego and in the few astral bits from the last life which managed to remain attached. However, in the following life, most people will never be able to access it. Apart from a few exceptional cases, it is only by working on oneself that one can reach deep enough to unveil the awareness of the Ego and the memories in its close periphery. Otherwise one is aware of only the superficial parts of the astral body, those that were collected just before incarnating and developed in the new life.

Now if most of our characters fall off as fragments after death, why do we keep so many characters from one incarnation to another? Structurally, a character is imprinted on a chunk of astral body. It develops gradually, as the imprints of more and more life experiences are added around a central core. In a very simplified manner, we could picture the character as consisting of a central core, close to the Ego, and of agglomerated astral material.

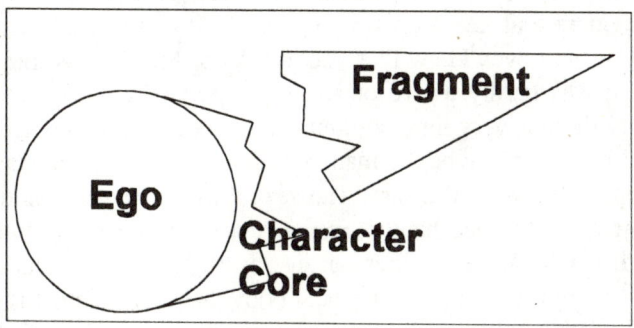

At death, the fragment falls off while the 'character's core' remains attached to the Self, with certain intense imprints collected in that life. In the following life, the character's core starts generating a full character around it again, if life circumstances allow it. The same applies to all characters.

Of course, all of this is presented in a simplified way. In reality many other subtle mechanisms play a part in the fascinating journey that leads from one death to a new birth.

CHAPTER 5

FRAGMENTS (continued)

5.1 Astral fragments

Following the description of the shattering of the astral body after death, the next step is to understand what happens to the astral fragments that are released. The main clue is that a number of these fragments remain endowed with a strong dynamism.

Let us consider, for instance, a part of the astral body that was involved in sexual activities. After death, the astral piece breaks off and becomes a fragment that keeps on wanting sex. It doesn't care whether you are dead. It doesn't even know that you are dead. Anyway, as long as you were alive, it was hardly aware of any other part of yours — remember that in reality there is very little communication between the different parts of the astral body. Or maybe the main action of the other parts was just to limit and repress the sexual desire. The sexual astral fragment is now in a position where, on the one hand it is not repressed any more, but on the other hand it is not satisfied either because it can't enjoy sex through you any more — there is no longer a physical body. So this fragment is floating in the astral space, unrepressed and craving satisfaction intensely.

The same can be said of the pianist fragment. All it wants is to play the piano, but there is no longer a piano or even hands with which to play. The fragment goes wandering around in the astral space, looking for something that could satisfy it. The same mechanism can be extended to all the fragments: each yearns for what it has known and enjoyed while you were alive. Each starts wandering in the space, looking for the satisfaction which it cannot get through the living person any more.

Now suppose there was a part of yourself that intensely enjoyed drinking beer. After death, the corresponding astral fragment floats in the space, deprived of what it wants, yet desperately desiring it. In the astral worlds, like attracts like. Therefore the beer-drinker fragment will naturally be attracted to (living) people who drink alcohol. For instance, the beer-

Fragments (continued)

drinker fragment may well find its way to a pub, or to any place where a group of people are drinking. While people are drinking, a certain thrill takes place in their astral body. The part of themselves which is thirsty for alcohol gets excited and vibrant when the fiery liquid is poured down their throat. This vibration can be felt far away in the astral space, and it attracts the floating beer-drinker fragment by a kind of magnetism.

Once the beer-drinker fragment gets close to the drinkers, it recognises their enjoyment, and even takes part in it. In the astral layer, things and forces are not as separate as they are in the physical world. For instance one does not need to be clairvoyant to know that emotions can be communicated from one person to another, even without words. Just by standing close to someone overanxious or fearful, one sometimes starts feeling the same emotion. In terms of subtle bodies, this has to do with a communication of the emotional wave directly from astral body to astral body. Similarly, the vibration of enjoyment generated by the drinkers is communicated to the beer-drinker fragment, through a kind of resonance. The closer to a drinker the fragment can get, the more intimately it will take part in the enjoyment vibrations.

Note that there is nothing evil or demoniacal about this beer-drinker fragment. What is described here is quite mechanical and has little if anything to do with moral concepts. The fragment is not moved by any conscious desire to harm. It is just a stereotyped piece of astral matter that seeks to repeat endlessly what it has been conditioned to do. So it will naturally try to stick to a place where lots of people drink, or to one particular person who drinks a lot. Suppose the fragment could find a way to take permanent residence around or inside the astral body of a heavy drinker, that would be ideal—from the fragment's point of view, of course! From the person's point of view, it would mean catching an entity.

This is the first type of entity: a fragment, issued from the shattered astral body of a dead person. Not all entities are fragments, as we will see later, but a large proportion of them are.

Their mechanisms are simple. Each astral fragment is very much polarised in one particular direction. All it wants is to repeat what is imprinted in it. When floating in the space, fragments seek what is familiar to them. Consequently, they will often be attracted to places where the dead used to live or work, and to people of the same family. For instance if a man has made love to a woman for years, it is quite logical that after his death, his sexual astral fragment or fragments may be attracted to the same woman. Even if such a simple and direct connection does not take place, the various fragments will often recognise some kind of familiar vibration among the people of the same family, like an inviting atmosphere. It follows that when someone dies in a family, the close relatives are those who run the greatest risk of catching one of the fragments.

Imagine the case of a mother who dies prematurely, leaving a young child behind her. The love for a child is one of the most intense emotions a man or a woman can feel. This love therefore involves large parts of the parents' astral body. Seeing oneself in the desperate position of having to die and leave the child motherless can only add intensity to the astral imprints. Once the mother is dead and her astral body is shattered, where else could the corresponding astral fragments try to go but to the child?

Of course life circumstances are not always that tragic. Still, it is quite common for parents or grandparents to have a high degree of emotional investment in their children. This makes the latter especially vulnerable to being parasitised by an astral fragment at the death of a relative, regardless of the age at which the relative dies.

5.2 An etheric layer around certain astral fragments

Many of the astral fragments which are released after death are actually 'coated' with a layer of etheric energy. The shattering of the astral body takes place in the days after death, when the astral body is still close to the disintegrating etheric body. For after death, the dead do not immediately make their way to distant astral spaces. They remain close to their physical body for a few days, and then gradually drift away. The dissolution of the etheric body does not happen in the hours following death, it also takes a few days. Thus in the days that follow death, while the physical body starts to rot, the etheric body dissolves and the astral body is shattered.

During life, the etheric and the astral bodies are closely interconnected. The astral body penetrates the etheric body like a hand penetrates a glove (but it also extends beyond the limits of the etheric). The various parts of the astral body tend to be more specifically connected with a particular area in the etheric body. For instance, the pianist astral part has a close connection with the etheric of the hands. The astral parts related to sexual enjoyment have a direct connection with the etheric of the genital organs, etc.

The dissolution of the etheric body takes place more or less simultaneously with the shattering of the astral body. Therefore, once released, it is not uncommon for astral fragments to retain the etheric part with which they used to be connected. Thus the pianist astral fragment may keep the etheric of the hands around itself and take it away when drifting in the space. Without the pianist astral fragment, the etheric of the hands would just disintegrate into etheric dust. Somehow the astral fragment acts as a backbone around which the etheric can remain agglomerated.

Due to the same mechanism, many of the astral fragments, especially the more structured ones, become surrounded by a layer of etheric substance. This will have a number of important consequences.

Firstly, it 'densifies' the astral fragment, it adds 'substance' and structure to it. Instead of being just an astral bit, the fragment becomes a more complex two-layered 'thing' (etheric-astral), with permanent exchanges taking place between the two layers. This increased degree of organisation significantly reinforces the fragment. It adds to its coherence and makes it much more resistant.

Secondly, the fragment will be able to use this etheric coating as a kind of glue, to attach itself to people or other living things (plants and animals also have an etheric body).

Another important consequence is that the fragment will have to feed its etheric layer. The etheric is the layer of life force, and it needs to be recharged to survive. Altogether, we can look at a fragment as a kind of two-layered autonomous 'creature', with its own desires and emotions, and its own movements in the astral space. However, the fragments are not fully developed creatures, they do not have all that is needed to start an independent life of their own. In particular the structure of their etheric is insufficient to be able to recharge itself. Therefore the fragments are only semi-autonomous. They will have to parasitise another living system to be able to get their life force. This explains why so many clients who are afflicted with an entity say that it drains their energy, sucks their life force, etc.

Note that this mechanism does not necessarily apply to all entities. As we said before, not all entities are fragments fallen from the astral bodies of deceased people. Moreover, not all astral fragments have an etheric layer around them, and those which have an etheric layer don't all have the same life force requirements and demands.

5.3 *The pros and cons of cremation*
The concept that fragments of the dead can remain and parasitise the living can be found in several cultures. In India, for instance, the custom of burning corpses is explained as a way of getting rid of these fragments.[1]

---

[1]Cremation is not found only in the Hindu tradition. The only way of disposing of the dead mentioned by Homer is by burning them. One also finds cremation customs among Persian, German and Slav ancient cultures.

This may at first seem paradoxical, for fire appears to belong to the physical world and therefore cannot touch the fragments made of astral substance. The Indian viewpoint is based on a different perception that sees fire not only as a physical process, but one that also involves intense etheric forces. Burning physical substances not only turns them into ashes but, to a certain extent, scatters the etheric attached to them. Therefore by burning the corpse, one achieves a certain degree of purification of the deceased's etheric body, turning it into a kind of etheric equivalent of ashes. Such a destruction of the etheric body by fire is not necessarily complete, nonetheless, fire is regarded as having a definite cleansing effect. The fragments of the shattered astral body can no longer surround themselves with parts of the etheric body, since the latter has more or less disintegrated into etheric dust. The consequence is that the astral fragments cannot 'solidify' themselves. They cannot consolidate their structure by coating themselves with an etheric layer, and so they remain much weaker. This significantly diminishes their chances of persisting as coherent astral bits instead of crumbling into undifferentiated astral dust. Even if they can maintain their astral structure, having no etheric layer makes it more difficult for them to attach themselves to living creatures and parasitise them.

Cremation, however, is not without disadvantages. If it is implemented soon after death, as it is in India, it has quite a traumatic effect on the souls of the dead. Just after death, the dead remain close to their physical body. It is a critical phase, during which some extremely important processes take place. The dead reexperience their life as a kind of panoramic motion picture. During this phase, the quality of energy in their consciousness will be essential in determining the type of astral plane to which they will be attracted. For in the astral worlds, like attracts like. If your consciousness (Ego + astral body) is serene and devoid of fear, you journey smoothly and in pleasant or neutral layers. If you are agitated by emotions, you become the prey of harassing forms and you drift into restless astral worlds.

In our civilisation, death has become a taboo that most people are reluctant to discuss. Moreover, our materialistic culture leaves individuals totally unprepared for death, believing that all existence stops with the end of the physical body. Consequently, most dead people do not understand what is happening to them at all. Actually many of them do not even realise that they are dead! Picture people who have just died and are still attached to their body, and who are not too sure of what is going on. If within a few hours, they are taken to a crematorium and their corpse is burnt, the result is likely to be panic. The dead feel themselves forcibly projected out of their physical body, terrified by the fire and stripped like a snail from its shell. This happens precisely at the time when they are in need of serenity so they may integrate their life panorama and take the right angle for their journey into the astral worlds.

So cremation has its pros and cons. On the one hand, it eliminates large quantities of entities, which will avoid an incalculable number of diseases and various other troubles in the population. On the other hand, cremating too early is a dramatic trauma for the dead, and is likely to disturb the critical early phases of their journey.

A solution could be to burn the corpse only when the life panorama is completed, once the dead person starts moving away. The duration of the life panorama most often indicated by occultists is around three or three and a half days. However, after three and a half days several etheric bits might already have been released into the atmosphere. I am not sure if there is an easy answer to the question of finding the ideal time for cremation. Rather than allotting a fixed time, it seems preferable for someone to keep in contact with the dead clairvoyantly and determine the favourable moment in each separate case. Also a qualified entity clearer should be in charge of detecting and eliminating the astral 'fallout'.

5.4 *Mourning customs in India*

Many of the traditional mourning customs and rites of the Indian Brahmans were guided by an understanding of the mechanisms of astral shattering, and aimed at protecting both the family and the rest of the community from being parasitised.

The first point to note is that, in India, the colour for funerals is white, not black. If a blue surface appears blue to our eyes, it is because it reflects a particular frequency of the visible spectrum. Black, on the contrary, does not reflect but holds; it is an absorbing principle. White reflects all the visible frequencies, hence the symbolism of purity associated with it: white is a principle that sends back everything — nothing remains attached to it. When a close relative has just died and there is a high risk of being contaminated by an astral fragment, traditional Indian people instinctively tend to wear white, to minimise the chance of negative energies getting attached. One can probably see an extension of the same atavistic wisdom in the fact that members of medical professions wear white in hospitals, or that fridges are usually white.

From this point of view, black is obviously not a safe colour to wear at funerals or when mourning the dead. Being an absorbing factor for vibrations of all kinds, it maximises the chance of catching the negative energies released by the dead, not to mention all those coming from the emotions of the family.

After a death in a traditional Indian family, all relatives are considered impure for a duration of thirteen days. In Sanskrit, this state of impurity is called *mṛtaka sūtaka*. Interestingly, a birth is also associated with a period of *sūtaka*, or ritual impurity (which will be discussed further in Chapters 6 and 7). However, the *sūtaka* of death is far more intense than that of birth. It is accompanied by several rites and cleansing practices which are intended to protect the family members and the community from astral 'fallout' from the dead. The degree of impurity varies depending on how closely related the deceased was. A thirteen day impurity time applies to directly related family members such as children and grandchildren. More remote relatives, up to the seventh degree, are said to be impure for ten days only, and relatives up to the sixteenth degree for three days.

After the death of a close relative, Brahmans must stop their work and activities for a few days. Family members are not allowed to take part in the usual ritual activities and worship of their caste. Moreover, the statue of the *iṣṭā devatā* (the deity with whom the family cultivates a special connection) must be removed from the house. On the day of the death, a friend who is not a member of the family, and therefore is not touched by the *sūtaka* or impurity, comes to the house and takes the statue away. It is brought back only on the thirteenth day, after the family members have gone through a number of purification rituals and baths. To understand these restrictions fully, one must know that Hindus consider that if they worship a god whilst in a state of impurity or uncleanliness, the result may well be a curse instead of a blessing.

The corpse is taken to the cremation ground and burnt as soon as possible. Corpses are never transported at night, when it is considered that there is more risk of being negatively influenced by elemental energies or spirits. Before the cremation and in the days following it, dumplings of wheat and water, called *piṇḍa*, are regularly offered to the dead during rites. After the cremation, dumplings are even offered to the pieces of bones that were not completely burnt.

As we saw with the *Kuei* and the customs of the ancient Chinese, it is not only in the Hindu tradition that one finds rituals and food offerings to disincarnate entities. The Tibetans too, have various rituals in which food and alcohol are offered and left unattended in plates and bowls, with the intention of satisfying the 'wandering souls'. Similarly, in ancient Greece it was common to make ritual offerings of wine, oil and honey to the souls of the dead. I will come back to this in Chapter 15, when dealing with the

Fragments (continued)

clearing of entities. From what has already been described, we can easily understand the thinking behind such practices: hunger and thirst are some of the most common longings of the astral fragments.

During the thirteen day period of the *mṛtaka sūtaka*, or ritual impurity, the family members are supposed to avoid eating various types of food: bread made from wheat (and therefore yeast), milk, clarified butter, various types of grains, tea, sweets and their favourite dishes. The question of avoiding meat does not arise because traditional Brahmans are strictly vegetarian, but diminishing the intake of proteins is obviously a major part of these dietary recommendations. Moreover, during the *sūtaka* the family is not allowed to feed visitors.

These dietary restrictions can easily be related to what we described earlier about astral fragments. Not indulging in tasty or heavy foods limits the chances of resonance taking place between astral fragments and family members. Many fragments would automatically be attracted if the family members were to eat gluttonously or indulge in any way. Apart from food, restrictions are imposed on sense enjoyment of various kinds. For instance the family members are not allowed to listen to music. Listening to music, in itself, would not attract many fragments. However, the Hindus consider that any form of enjoyment creates an excitement of the whole astral body, which in turn accentuates all emotions and desires. Therefore not listening to music can be understood as another way of keeping the astral body as quiet as possible, which follows the same logic as the plain, boring diet.

On the tenth day, everything and everyone in the house is washed. Men shave their moustache, and cow dung is applied to the walls of the house. In India, cows are regarded as a highly purifying agent. All ashrams, for instance, try to keep at least one cow. Cows are not only a spiritual blessing (Krishna's paradise is full of them), they are also said to absorb negative energies. Negative vibrations are said to be neutralised by cows just as snow flakes dissolve in a cauldron of boiling water. So one can easily understand why traditional Indian families, if they could afford it, used to have a cow milked daily for thirteen days on the precise spot where the corpse had been cremated: the more *bhūtas* (etheric and astral fragments) the cow absorbed, the better for the family.[1]

---

[1] The word *bhūta* is the common Sanskrit word for entity. Interestingly, it is constructed exactly the same way as the word entity. Entity comes from the Latin *ens* and the verb *esse*. In Sanskrit, *bhūta* comes from the verb *bhū*, to be. Like the Latin word *ens*, *bhūta* means 'that which is', a being. The fact that *bhūta* is commonly used in Sanskrit to mean disincarnate spirit

Entities

In India, great yogis are not cremated but buried. The reason is that enlightened yogis are supposed to have purified and unified their astral body, so no passions or desires are left in them. Ideally their astral body is not shattered after death, and they take it with them into the spiritual worlds. No astral fragments are released, so there is no reason for cremation. On the contrary one wishes to benefit as much as possible from the good vibrations accumulated in their body due to their spiritual practices. For this reason they are buried, traditionally, in the vertical position, and their tomb becomes a place of inspiration.

Interestingly, infants under the age of eighteen months are not cremated either, but buried. Eighteen months corresponds roughly to the time when children utter their first words, thereby connecting with different aspects of the mind. This is also the period when young children lose a lot of their psychic abilities.[1]

Before leaving the Hindu customs, I must emphasise that the duration of thirteen days should not necessarily be taken literally. If you read 'thirteen days' in a twentieth-century western treatise of physiology, it means thirteen days, or three hundred and twelve hours, and nothing else. If you read 'thirteen days' in a Sanskrit treatise that was written hundreds, if not thousands of years ago, then what is meant is: 'a certain number of days, related to the symbolic meaning of the number thirteen'.[2] Too literal an understanding of figures found in ancient Sanskrit texts causes gross misunderstandings, and is due to a lack of familiarity with the Indian culture.

5.5 Mourning rules in the Jewish tradition

The understanding of fragments sheds a different light on a number of restrictions imposed on mourners by the Code of Jewish Law.

For the first period of seven days, called *shiva*, mourners are not allowed to study the Torah, which parallels the Hindus' restrictions on worship. During this first week, the mourners are forbidden to perform any work. Sexual intercourse, kissing or embracing is forbidden. One is not

---

is another good reason to use the word 'entity' with the same meaning in English.

[1] See Chapter 19, 'Baby Work' in *Awakening the Third Eye*, by the same author.

[2] In the western tradition too, one finds many associations between the number 13 and death, as in the thirteenth arcanum of the Tarot deck, for instance.

allowed to sit or sleep on a bench or on pillows and cushions, one must sit and sleep on the ground.

A second group of restrictions concerns a longer period, called *sheloshim*. The *sheloshim* lasts thirty days in the case of the loss of a relative, and twelve months after the loss of one's father or mother. During that period, mourners are not allowed to join any feast or party. They must not invite friends, accept invitations, send or receive presents. Haircutting, shaving and nailcutting are prohibited. Moreover, the mourners must change their seat at the synagogue (as if to avoid being located by a fragment!)

5.6 A note on Alzheimer's disease and schizophrenia

While discussing the topic of fragments, it may be of interest to share certain observations I have made of patients suffering from Alzheimer's disease and schizophrenia.

Alzheimer's disease is a terrible degenerative disease of the nervous system: the patient's brain becomes shrunken and atrophic. It usually starts in the last decades of life with insidiously minor troubles such as memory deficiency or changes of behaviour. It gradually leads to a premature state of senile dementia, in other words mental disintegration.

I have dealt with clients who had a parent suffering from advanced stages of Alzheimer's disease. In some of these cases, I have observed without any possible doubt that the clients were catching fragments from their diseased parent, even though the latter was still alive. They were being parasitised by fragments which had broken off their parent's astral body. This has led me to think that in Alzheimer's disease, at least part of the astral body's shattering which normally takes place after death, happens while the patient is still alive, during the period of mental disintegration.

Schizophrenia, a no less terrible disease, often strikes young people. While dealing with schizophrenic patients, some of their so-called hallucinations have appeared to be genuine non-physical perceptions. However, they take place in a hectic and uncontrolled way that generates high anxiety, and in the context of a major disorganisation of the personality.

We have discussed earlier how the consciousness of the astral body is reverberated into the physical nervous system as physical mental consciousness. We saw how it is the physical mental consciousness that gives the illusory impression of being one single person instead of an astral mob. In the case of schizophrenics, the physical mental consciousness is deeply altered. Schizophrenics are projected against their will into a panoramic

Entities

vision of their astral body. They start perceiving themselves as the mob, which in a way could be regarded as a spiritual achievement. But this experience comes too soon, without the light of the Self, and in the context of major psychiatric disorders.

As with Alzheimer's disease, I have observed certain clients who had a close relative suffering from schizophrenia, and who were parasitised by a fragment coming from this relative. This has led me to suppose that the disintegration of the personality which takes place in the more advanced stages of schizophrenia is accompanied by (or perhaps due to) a premature shattering of the astral body.

CHAPTER 6

ENTITIES, PREGNANCY AND GYNAECOLOGY

6.1 Entities and female reproductive organs

Among the most frequent locations described by female clients when exploring entities are: in or around the ovaries, the uterus and the vagina. In certain cases, the presence of the entity is associated with gynaecological disorders, ranging from premenstrual tension and irregular menstruation to malignant tumours. However, in many other cases, there is no physical illness or symptom, just an entity attached to the reproductive organs and creating the usual syndrome comprised of cravings, energy depletion, and various psychological troubles, as described in Chapters 1 and 2.

This location in the reproductive organs has no parallel in male clients. Even though entities often interfere with sexual desire and libido in general, it is actually quite rare for men to describe an entity located in the vicinity of their sexual organs.

Why should so many entities be attracted to the female sexual organs?

One reason could simply be a question of space. Of course, entities are not physical, but many of them are surrounded by a layer of etheric energy. If we look at the way the etheric body permeates the physical body, we discover that the densest physical tissues are much less penetrated by the etheric than the watery tissues are. For instance, the etheric body does not permeate the bones much. Physical density seems to make it more difficult for etheric energies to penetrate a tissue. It follows that it is easier for foreign etheric energies to find a place in empty body parts such as the natural cavities.

If we look at the various cavities of the body, we find that the heart chambers and the blood vessels are closed and extremely protected from the external world. The digestive tube, on the contrary, is open at the mouth and the anus, and one could certainly think of the stomach as an organ in which entities could try to nest. But in the mouth, the stomach and the small

Entities

intestine, the intense chemical operations of digestion are quite unfavourable for entities. During digestion it is not only the physical nutrients that are broken down, but also their etheric part. Consequently, if an entity happens to go astray in your gastro-intestinal tract, it is very likely to be 'digested'. The entity's etheric having crumbled under the fire of digestion, its astral component has little chance of maintaining itself inside you. It will either be naturally eliminated or crumble into astral dust.

There are exceptions to this rule. In particular, various perverse energies can manage to find their way inside your system via unclean foods (especially fats and meat). However, perverse energies and entities are not the same thing, and the probability of catching an entity through eating remains more than minute. The destructive effect of digestion on entities is even exploited by certain shamans who, to rid their client of an entity, suck it into their mouth and swallow it! I would certainly not recommend this as a method that any healer would be able to 'stomach', nor as the most elegant way of clearing entities, but the fact that it is possible has interesting repercussions on the way we understand entities. In particular, it clearly shows that the stomach isn't a suitable migration place for entities.

The situation is quite different in the lower colon. As far as digestion is concerned, not much happens in the colon—mainly the reabsorption of water. There are no digestive secretions full of enzymes, no aggressive digestion processes. Therefore in the descending and sigmoid colon one finds another natural cavity in which it seems logical for entities to seek refuge. This could explain why so many clients, males and females, point to the left iliac area as the location of their entity.

Apart from the space factor, what other reasons may account for the number of entities found in female reproductive organs? The fact that the vaginal cavity is open to the outside probably makes it easier for foreign energies, including entities, to penetrate. Besides this, many entities will obviously be attracted to this area because of the concentrated life energy that is associated with reproduction. As we have seen before, most entities crave vital energy. A certain quintessence of the life force is stored in the sexual glands.

The left iliac area is not far from the female reproductive organs, which is probably another reason why it is one of the most common locations for entities. Note that in traditional Chinese medicine, there is an acupuncture point called *Qi Hai* (Conception 6). This point is situated approximately one inch below the navel, in the same area as the Hara, in which practitioners of martial arts learn to concentrate energies. *Hai* means sea, and *Qi* is the Chinese word for etheric energy, or life force. The area

Pregnancy and Gynaecology

of *Qi Hai* and of the Hara can be regarded as a powerhouse of the life force in the body. Since most entities crave life force, it is understandable that they will often attach themselves somewhere in the lower abdomen, in both men and women.

Another factor that attracts entities to female reproductive organs is that many entities have an obvious interest in sex. It is quite common for clients to describe that their entity gets a thrill out of sex due to the emotional intensity and of the release of life force that comes with intercourse, and due to the concentrated life energy associated with the semen and the ovaries. By being attached to reproductive organs, entities have a choice place to enjoy and drain the etheric energies that are exchanged and released during the sexual act.

It also appears that many entities have a need to feel safe and protected. As will be seen in some of the examples given below, it is not rare for women who carry an entity in that area to subconsciously adopt a caring and fostering attitude, as if the entity were their baby. The motherly nurturing makes the entity feel snug and secure, and taken care of. Such a situation is particularly satisfying for the many entities that just want to slumber in darkness and not be disturbed, like a child wishing to stay in the womb forever instead of facing the painful complications of 'going outside'.

There is another factor, which might appeal to psychoanalysts. Wandering souls in the astral, whether human or animal, have a fascination for the womb. While travelling in the astral in between two reincarnations, there are stages where souls feel a huge craving for the security and other benefits that come from being in a matrix. They are moved by a burning desire, an intense 'itch' imprinted in their astral body, which pushes them to rush into a mother. Imagine the thirst of people who have walked for days in the desert without any water, and who suddenly reach a pond. What else can they do but run and drink? It is with a similar frantic intensity that most souls dash into their mother's womb.

As we will see later, apart from perhaps a few very rare cases, entities are not full spirits which have lost their way, but just etheric and astral fragments or energies. Still, part of them is made of astral substance, similar to that of the astral body of souls travelling in between two lives. The craving for incarnation into a matrix is deeply stamped in that substance, just as sexual desire is deeply imprinted in incarnated human beings and animals. This archetypal fascination for the womb is a powerful driving force, and one of the reasons why so many entities are attracted to female reproductive organs.

Entities

Now, before looking at a few examples, let me say again that entities are *not* everywhere! No-one would worry about cholera in the case of a simple diarrhoea, or about brain tumour in the case of an ordinary headache. Similarly, it would be absurd for a woman to start worrying about entities each time she gets premenstrual tension. Just because a significant proportion of female clients with an entity discover that it is located in or around their reproductive organs does *not* suggest that all women have entities! This cannot be emphasised enough, because a few people, once they learn about entities, tend to suspect them everywhere and develop a kind of spook-phobia which is completely unjustified.

Case study 6.1 Forty-two year old woman. She comes to consult me about a fibroid. The tumour was discovered five years ago, and has now reached the size of an 18 week foetus. The fibroid is bleeding more and more, and the client's surgeon has decided to operate and perform a hysterectomy.

At the beginning of the ISIS session the client has contacted a darker and denser patch in her energy, in the area where she can feel the fibroid.

Is there any emotion or feeling that can be related to it? –Fear. Grief.

Does it have a shape? –Yes. It is like a pear with the narrow part at the bottom. There is something on the outside of it. Like a growth, a hardness. The hardness is just under the skin. I'm thinking of these longer pears that are quite willowy and beautifully shaped.

What does 'it' want? –Comfort. It has to do with violence and loss and abuse.

What kind of violence? –Rape, when I was ten. The man did not do it unkindly but I was too young. He was my stepfather. He had a relationship with me for five years. He should not have done that to me but nevertheless I loved him very much. When I went to boarding school I told one of the nuns. Then my mother learned about it and left him. And he killed himself. I loved him very much. He was very tormented. He was a doctor. When he died, it was a time of madness for me. I was not allowed to cry in front of my mother because she hated him and she was happy he was dead. I felt like his ally. He was brutal and violent to her, and she made herself his victim. But he had always been nice to me, except that he abused me sexually.

Is there any connection between him and the fibroid? –Yes. His despair. It's just rage and despair. Actually, now I can feel his presence in the fibroid. [Crying:] I thought I had finished with that.

Pregnancy and Gynaecology

What does it want? –Comfort. Same as when I was a child and he [the stepfather] was trying to get comfort from me. It wants me to give it life, and I do, by having it there.
Are there some foods that it enjoys? –Meat, blood. It likes meat.
What happens to the presence when you eat meat? –It's strengthened. I very rarely have meat but sometimes I want it so badly that I have to eat some immediately. He [the presence] wants meat. It's connected with my own bleeding too, like a vicious circle: when I eat this bloody food, it's easier for the presence to make me bleed.
What happens to the presence when you bleed? –I have the feeling of a certain malevolence. It takes some pleasure out of my bleeding. It's like my stepfather's sperm stayed there and grew out of my bleeding.

If we apply the pattern developed in the last chapters to this case study, we can understand this entity as being an astral fragment of the stepfather. The fact that the client recognises the stepfather's presence in the fibroid is of course the first important factor suggesting the presence of a fragment. Moreover, the circumstances of the stepfather's death, a suicide in tragic circumstances, also suggest that the fragments released from his astral body must have been endowed with a particularly strong dynamism. The 'time of madness' the little girl went through, unable to express her emotions openly, certainly was a facilitating factor for the fragment to invade her. The love she felt for her stepfather was also a factor that allowed the fragment to come in.

As soon as the entity was cleared, the fibroid started diminishing in size. After a few days, it could no longer be felt during sexual intercourse. With a few other complementary healing techniques, the fibroid kept on diminishing gradually. Surgery was avoided, and the client saved her uterus.

6.2 When does the baby's soul incarnate in the womb?

While exploring entities, I have found that a significant number of female clients see in their entity the continuation of an embryo that had been aborted, sometimes as long as fifty years before. From a materialistic point of view, once the embryo has been removed by curettage, the abortion is completed. However, from the point of view of subtle bodies, an embryo is not just physical matter. The physical embryo is but a fraction of the whole embryo, which is also comprised of subtle parts. Removing the physical part is not necessarily enough to remove its subtle parts. Whatever etheric and astral energies are left after an abortion are quite likely to become an entity.

In terms of subtle bodies, what exactly is an embryo made of? The answer depends on the stage of the pregnancy. Taken at the moment when the spermatozoid penetrates the ovum, the egg is made of the physical cells plus an extraordinarily concentrated etheric energy. In the last part of the pregnancy, the foetus is comprised of the four layers: physical, etheric, astral and Ego. These four envelopes are not functioning in the same manner as in an adult, but still they are present. The critical question is, of course, when does the baby's soul (or more precisely the baby's astral body and Higher Self) incarnate into the mother's womb?[1]

Different spiritual traditions have given different answers to this question. For instance, the Hindu tradition considers that it is in the fourth month of pregnancy that the baby incarnates, meaning that the baby's astral body and Higher Self arrive in the mother and become attached to the physical and etheric embryo.[2] According to Ayurveda, or traditional Indian medicine, in the fourth month the heart of the foetus is well developed, which allows the baby's Higher Self to take position in it. From then on, the mother is called *dauhṛdinī*, 'the one with two hearts', and due to the baby's desires she starts feeling cravings.

This may appear to offer an easy answer to the problem of abortion: from the point of view of Ayurveda, before the fourth month the embryo is just a piece of flesh with a bit of prana (life force) in it, exactly like a vegetable. There is therefore no spiritual contra-indication to abortion, no offence against a higher form of life.

To the objection "But this embryo is alive. If you kill it, you destroy a life", I remember hearing a great Indian master answer: "Until the fourth month, the embryo is no more alive than a tomato. If you think like this then you should stop eating tomatoes because by eating vegetables you destroy a life in exactly the same way." [3]

This opinion is supported by a large number of Indian masters, and backed by the authority of ancient Sanskrit texts. I therefore mention it

---

[1] For reasons developed elsewhere, I prefer to restrict the use of the word *soul* to the astral body.

[2] See for instance *Suśruta-Saṃhitā, Śārīrasthāna*, Chapter III, verses 14-30. A translation of this passage with commentary will be found in the chapters on death in *Tantra, Body and Worlds,* by the same author.

[3] Swami Satyananda Saraswati in a Satsang given in 1978 in his ashram in Monghyr, Bihar, India. How interesting that he chose tomatoes for his example!

whenever I deal with a patient who has made up her mind to have an abortion. Having to terminate a pregnancy often creates immense sorrow and guilt in a woman, sometimes with devastating and long-lasting psychological consequences. Seeing no benefit in so much guilt and grief, I welcome anything that can bring some relief. Experience has shown me that a number of women do experience a significant relief when pondering on the fact that, from the point of view of Indian wisdom, they are not destroying a life by having an abortion, provided it is performed before the fourth month of pregnancy.

Whether things are really that simple or not is another question. All esoteric traditions are far from agreeing on this point, some placing the arrival of the baby's soul much earlier than the fourth month. Moreover, whatever may have been the case two thousand years ago in India, we know very well that nowadays women do not wait till the fourth month of pregnancy to have cravings. So if we were to take the beginning of the cravings as an indicator, as Ayurveda does, it would mean that the baby's soul sometimes takes position inside the womb as early as the first month.

When dealing with subtle bodies, it is important not to be too mechanistic. In particular, it would be quite illusory to picture the four layers (physical, etheric, astral, Ego) as four Russian dolls, one inside the other. This applies especially to a baby, even after it is born: its astral body and Ego are far from being tightly bound 'inside' the physical and etheric bodies. A baby can hardly remain awake more than a few hours. During sleep, the astral and the Ego drift away from the physical and the etheric. The fact that the baby falls asleep all the time indicates that its astral body and Ego have great difficulty remaining inside the physical and etheric bodies. In other words, for a good year after birth, babies are fantastic astral travellers and spend most of their time out of their body, floating in high astral spaces—hence the great spiritual inspiration one can draw from the company of a baby.[1]

If even after birth babies are hardly present in their body, then obviously they are even less so during pregnancy. From the point of view of subtle bodies, incarnation on Earth does not happen overnight. There is not one particular day during pregnancy when the mother can say: "That's it. The baby has arrived inside!" The incarnation of the astral and the Ego into the physical and etheric is a very gradual process. It is only around puberty that the astral will start taking its final position inside the physical body, and the process will not be completed before adulthood! When esotericists

---

[1]See Chapter 19, 'Baby Work' in *Awakening the Third Eye,* by the same author.

say that astrological maturity is reached around the twenty-eighth year, they refer to the fact that for most people it takes that long for the Ego to arrive fully on Earth. Of course, in reality there is no fixed time. Certain strong spiritual souls are quite fully incarnated as early as puberty or even before it. On the opposite side of the scale, billions of people on Earth will never incarnate much of their Ego. They may perform all their daily activities frantically and show a facade of responsible presence, but their Higher Self is thousands of astral light-years away. All this shows one must cultivate fluid thinking if one wants to comprehend the laws of life, and in particular those related to subtle bodies.

In terms of subtle bodies, if one looks at how a baby 'incarnates' in its mother during pregnancy, what does one see? In the very beginning the embryo is just a bit of physical substance permeated by the most gorgeous, golden and heart-warming etheric energies. That is the precursor of the baby's lower complex, meaning its physical and etheric bodies.

Then at some stage, a soul (astral body + Ego) comes around. In the beginning, this is nothing more than a nearness, it can hardly be perceived, either clairvoyantly or intuitively. Then progressively, the soul gets hooked into the lower complex (physical + etheric). This is a very gradual process that takes place over weeks, if not months. Moreover, it varies quite a lot from one mother to another. So it appears to me somehow artificial to try to establish a fixed incarnation time and say: before that point the baby is not here, after that point it is in the mother's womb. In the early stages especially, the link between the mother and the baby's soul is fragile and can easily be broken, which would result in the baby drifting away and the pregnancy finishing with a miscarriage.

It would therefore be totally unrealistic to think that, after a termination, the whole of the baby's soul might remain trapped in the mother's womb. If one year after the delivery the baby can still hardly maintain its astral body inside its physical body, how could it be trapped inside the womb during pregnancy? As soon as an abortion or a miscarriage takes place, the link between mother and baby is broken and the baby's soul drifts away into the astral spaces.

6.3 Entities after miscarriages and abortions

The problem is that during the time of nearness, while the baby is gradually establishing its connection with the mother, parts of the baby's astral body become attached to the embryo in the womb, i.e. to the physical and etheric vehicles prepared for the baby by the mother. If a termination takes place, fragments will break off from the baby's astral body. This process is not nearly as drastic as the astral body's shattering after death. Still, a few astral bits drop off. Now if an abortion or a miscarriage takes place and if one only takes care of the physical side, leaving parts of the embryo's etheric inside the mother's womb, where will the astral fragments go? The answer is obvious: they will stick to their corresponding etheric parts in the womb, and the woman will have inherited an entity.

A number of factors tend to reinforce the attraction of the astral fragments to the womb. The immense emotional involvement of a mother towards her baby obviously plays a major part. Moreover, it is not by chance that a particular baby comes to a particular mother: there has to be an affinity, a polarity of some kind between them. All this contributes to cementing the fragments in the womb, and makes any termination a high risk situation for catching an entity. The fact that similar mechanisms may occur each time an egg is fecundated and an early miscarriage takes place is probably one more reason why such a large percentage of entities in female clients are found in the area of the reproductive organs.

***

Case study 6.2 Twenty-four year old woman, student. She had an abortion six weeks before this session. She went to the intervention in quite a serene mood. After this, she felt completely scattered, crying all the time without understanding why. Her menstruation did not come back. She displayed a moderate fever, between 38°C and 38.5°C, for which no cause could be found. Antibiotics were prescribed but did not improve the situation.
At the beginning of the ISIS session, a dark cloud associated with a presence was identified in the lower abdominal area.

What are you feeling? –It's in my uterus. I feel hostility coming from this area. It's black and it's angry at me. It's the embryo, it's stuck inside and it wants to punish me. Not only because of the termination, but for all the bad things I have done in my life. It feels dark, like a heavy blanket, and deceitful. It's like a terror, dark. I can now make the connection with the dark feelings, like I was having today. It feels like me being scared, but it's not me who is scared, it's the cloud.
Are there foods that it likes? –It seems to like sugar, for some reason. Lately, I've been wanting chocolate and sugary things much more than usual. It seems to be coming from there.
Why is it so angry with you? –There does not seem to be any reason. It's just very afraid and it manifests it with anger and by making me sick.

The dark anger coming from the fragment does not indicate that the baby would have been a monster. Such a fragment is nothing more than a broken-off bit of the baby's astral body. Because it separated from the baby's soul, the fragment assumed new characteristics it did not necessarily express as long as it was part of it. When they are left alone, unrepressed and unrelated to the whole, fragments often tend to manifest in a much darker way than when integrated around an Ego.

Of course some people will argue that any distressed woman after an abortion may well see angry black clouds in her uterus. But in this case, apart from the fact that the young woman could clearly see "something that went up, out of my uterus" during the clearing I performed, her temperature went back to normal in the following hours. She was immediately calmer, feeling definitely relieved. Her menstruation recommenced three days later.

Case study 6.3 Fifty-four year old woman, secretary. Here is the way she described what she felt when a fragment resulting from an abortion was cleared.

–I suddenly recalled the abortion I went through at the age of seventeen [i.e. 37 years ago]. I could justify the decision then and now, and had no regrets about it. But when I started feeling the presence of the soul that had been aborted, I had the sensation that things had been left unsaid. I felt the need to say that I was sorry for any problem I had created by terminating this life. It's like I was gently holding the soul close to my heart, surrounding it with a love that came from deep inside. I cried, and I said goodbye, something I had failed to do so many years ago. Then the soul was released upwards to the light. One week later, I had a period, for the first time in more than four years. This period lasted five days: good flow, good colour, as if I was seventeen. Eighteen months have passed since, and I have not had another period.

Case study 6.4 Forty-three year old successful businesswoman. Completely frigid for many years, she suffered from obstinate constipation.
In the ISIS state, we identified a 'darker, denser, heavier area' in the left side of the lower abdomen.
Are there any emotions or feelings associated with it? –It's like pressure, like being held down, not being able to move.

Pregnancy and Gynaecology

Could it be that there are some foods you eat that it enjoys? —Sugar and chocolate. Sweet things, cakes and pastries.

If it could speak through you, what would it say? —Don't come near me. Just let me be. It hurts. It's too heavy, it's too much pressure. It's like a baby. That's why it doesn't want me to have sex. Sex hurts. It's like a baby. [Crying:] It's the baby I did not have. I had an abortion when I was twenty-two. I was in England on my own. It still wants to be born. I always wanted to be a mother but I couldn't. I have a dog and he curls up there on my tummy all the time. The baby inside also wants me to hold my tummy. It's big. It needs space. It's about half the size of a baby. It's a little girl. She feels safe inside my tummy and she doesn't want to go anywhere. She did not want to be born, to go out. She just wanted to remain safe inside. She doesn't want me to have sex. My husband is big, it hurts the baby. [The client is caressing her belly as if she were pregnant.] She did not want me to have any other baby either, because there wouldn't have been any room left for her. She got scared when I first got married because my husband and I were talking of having a baby.

Did she play a role in your decision not to have a baby? [Crying] —Yes. When I wanted to have a baby, twenty-five years ago, she made me fall sick. My stomach got swollen and I got very tired and sick.

Could there be a part of yourself that's attached to her? —Oh Yes. I want to hold her. She's my baby, I feel a lot of love for her. [The client keeps on patting her abdomen:] I like my tummy. I can really feel this baby now. It's strong and firm.

You mean you can feel it physically? —Yes, firm. She's curled up in the foetal position. She's very comfortable. Her head is there; her feet are there. She doesn't like me to exercise, because she gets disturbed and woken up. I always have to fight to do exercise. Same feeling of tiredness as when I was pregnant. ['When' was a slip: the client had never been pregnant, apart from her abortion, and I later checked that she was not referring to that.]

Are there people around you that the baby likes or dislikes? —She likes my dog. She's happy when the dog curls up close to her. She doesn't like my husband because he is too big. Twelve years ago I got very attached to a man. Now I think the baby adored him. He was married. It was very difficult to break from this man. It took me months to recover.

What does the baby want? —She wants to sleep. She wants to be left alone and sleep. Each time I don't want to face something, I go to sleep. She doesn't want me to feel my body too much, because if I do she can't float as nicely. She hates it if I try to exercise. She's lazy, she doesn't want to

> get outside. I only go out when I have to. When I'm indoors I never want to go out.

In this case we see a fragment left over from an abortion, which had become a major factor of interference in the client's life. At the end of the session, she commented about all the problems she could have saved herself if the 'baby' could have been cleared just after the abortion.

To end this section, let it be clear that my perspective is not to take a position either in favour of or against abortion, but to describe certain mechanisms. From the point of view of subtle energies, the way abortions are presently conducted is incomplete. It leads not only to multiple psychological problems, but also to physical ones ranging from painful menstruation to malignant tumours. I am not suggesting one should stop conducting abortions but that if one does, they should be properly performed. The energy side of the operation should not be neglected.

I could fill an entire book with the case studies of clients who had a post-abortion entity. The corresponding disorders are sometimes negligible and sometimes tragic, but in every case I can't avoid thinking "What a waste!" Once you have found a skilled clearer, and I don't doubt there will be many in the future, clearing an entity is simple, quick and painless. If a proper clearing could have been performed in the weeks following the abortion, so much trouble could have been avoided. However, once a gynaecological problem has evolved for years around an entity and has developed into a major illness, it is no longer obvious that the clearing will suffice to resolve the problem. Although the clearing will help, past a certain degree of development a physical illness gains a momentum of its own. Thus, in many cases, it is not enough to clear the entity to heal the client, even if it is the entity that started the illness.

Let me therefore stress again that ideally, any woman who has gone through an abortion, a miscarriage or a delivery should be checked by someone who knows about energy and is qualified to clear entities. By doing this, innumerable troubles will be avoided.

CHAPTER 7

ENTITIES, PREGNANCY AND GYNAECOLOGY
(continued)

7.1 Entities and the placenta

In Latin, placenta means 'cake'. The placenta is the spongy vascular organ that surrounds and feeds the foetus in the uterus. As it is responsible for the life sustenance of the foetus, the placenta conveys not only oxygen and nutrients but also life force. In terms of subtle bodies, this means that the placenta is not only a physical structure. It also has an important etheric layer attached to it, through which the foetus receives life force from its mother during pregnancy.

After the baby's delivery (or after a late miscarriage), the placenta is released from the uterus. But once more, the placenta is not only made of physical substance, therefore its non-physical parts will have to be released too. Some of the etheric parts are expelled right at the beginning, together with the physical placenta. Others remain inside the womb and are eliminated naturally in the coming hours, days or weeks.

However, it sometimes happens that etheric bits of the placenta remain stuck inside. They may either remain as bits of etheric and behave like a perverse energy, or attract an astral fragment and become an entity. In either case, it means potential health troubles for the mother, especially when one considers that the post-delivery time is often accompanied by a great emptiness of energy. The fragments or perverse energies may create or accentuate a postnatal depression, or remain latent and silent for years before creating a physical illness.

Case study 7.1 Twenty-two year old woman, student. Two years before, she miscarried during the sixth month of pregnancy. Here is the way she reported the session.

–I lay with my eyes closed, relaxed, and began the ISIS technique with Samuel. My breath deepened and slowed.

–The first sensation I became aware of was a tight feeling in my solar plexus. The tightness spread down over my belly. My stomach felt stretched to the point of pain. I then felt twinges in my lower abdomen and near my ovaries.

–Suddenly, my belly went from stretched flat to really swollen. I could see how bloated I was. I had miscarried in the sixth month and my stomach was easily as large as it was then. The twinges developed into definite waves of cramps that intensified till they swept down my spine. I frowned, but really I wanted to cry out. The pains were exactly like labour pains and I was seized with panic, not understanding the process at all.

–My cervix seemed heavy and open, and I PHYSICALLY felt something begin to move down my vaginal canal. It was not a vague impression but a tangible physical sensation. I felt my mind call: "Oh my God! I'm going to miscarry again". I became caught in all the intensity of the mixed emotions of that time. It was like a physical replay of the labour, but this time mixed with the fear that I did not know what was moving out of my body. With each wave of cramp the mass moved further down.

–I remember feeling a strong heat. When the mass emerged I 'saw' what for a split second seemed to be an organ—like a liver. Suddenly I realised that it was a placenta, but much larger than the one I had given birth to physically, and it was very black.

–Though I had not opened my eyes I could see the mass between my legs or maybe just above them for a few moments. Then it became unclear, it seemed to fade. But I remember the exhaustion and a great sense of release and relief. My belly felt empty and light, a feeling which has remained with me, at varying intensities, since then.

Case 7.2 Thirty-six year old woman, salesperson, mother of two children. She had uterine cancer five years after the birth of her second child, and was treated with surgery and chemotherapy. Two years later, the cancer started again and was treated similarly, after which the client appeared to be healed for a while. When I saw her, eighteen months after the second

Pregnancy and Gynaecology (continued)

episode, her health had suddenly started deteriorating again. She needed increasing doses of codeine to fight her abdominal pains, and she could recognise in herself all the signs that had announced her two preceding episodes of cancer. At that point she was quite desperate, feeling life drifting away from her.

What are you feeling? –I can feel a sort of black mass where my uterus used to be.
About what size? –About one inch.
What kind of emotions or feelings could be associated with it? –It's foul, very violent. Wild. It wants passion, sex, raw sex. It likes violence. It likes it when I have pain, it feeds on it. And it likes codeine too.
What happens to it when you take codeine? –It gets stronger. It can control me more. It's wild. That wild vibration started after the birth of my second child. I remember I told the doctor I felt the placenta had not completely gone. He checked and he said everything was OK. But still I felt like something had stayed inside, and it was the same vibration—but now it has become much more intense. For the first few months after the birth I was trying not to feel it, pretending it wasn't there. So I went through a time of being incredibly prudish, which was not like me, to say the least. If anyone cracked a sexual joke around me, I would get furious. My husband hated me.
Is there a connection between this black mass and the cancers? –It IS the cancer. It feels exactly like how I felt when I had cancer. [Crying:] And now it wants to start it again. This thing is going to kill me.

The relief brought about by this clearing, which was accompanied by an intense course of ISIS sessions, was spectacular. Within a few days, the client literally changed colour, losing the gloomy greenish-grey hue of her skin. All her friends commented on how radiant she looked. The pain decreased, which allowed the dosage of codeine to be immediately reduced by half. Gradually she stopped using the drug altogether. In a matter of weeks, she went from dying to starting a new life. Five years later, she still appears completely healed.

I would not want any reader to get over-enthusiastic. Apart from a few exceptional cases, I am *not* suggesting that it is enough to clear entities to get rid of fatal diseases. Firstly, not all cancers and diseases are due to entities. Secondly, clearing entities early enough can avoid a number of diseases, but once a physical illness is solidly established, the situation is quite different. If an entity has originally been responsible for the illness, then a clearing will make the cure easier and decrease the chance of recurrence. However, in most cases the client will not be healed just by clearing the entity, for physical disorders tend to acquire a dynamism of their own with time, and may therefore persist after their original cause has been removed.

Clearing entities should be part of preventive rather than curative medicine. The weeks after a delivery are a critical time when a 'checking' is needed, especially (but not only) if there is any suspicion that a non-physical part of the placenta may have been retained.

7.2 Astral transfers during pregnancy

Let us now consider certain extraordinary mechanisms related to pregnancy. Mother, the French born yogi who was for many years in charge of the Sri Aurobindo Ashram in Pondicherry, India, commented in her *Agenda* on strange connections between parents and children.[1] Her vision was that during pregnancy, a number of the mother's repressed psychological tendencies tend to 'come out' and contaminate the baby. All sorts of vicious instincts and negative qualities that the woman had buried inside her subconscious and forgotten about are released, and pass into the child. Mother, who was ninety-one at the time, remembered something she had read 'very, very long ago', possibly by the French author Renan: "Beware of the children of nice, respectable parents because birth is a purge". (Mother was laughing while saying this.) The same French author suggested one should also watch the children of really 'bad' people carefully, because these children too often show dispositions which are diametrically opposed to those of their parents.

Mother was puzzled by the horrible temper of the children of certain very well-to-do parents in her ashram. So she applied her immense vision to childbirth, and came to the conclusion that having children was for many people a way of throwing out everything they didn't want in themselves, exactly like a purge. Seen from this perspective, the common saying that pregnancy heals many of the mother's troubles takes on a new dimension.

In terms of subtle bodies, this mechanism corresponds to a transfer of certain parts of the mother's astral body into that of the foetus. This transfer does not happen overnight. It is a gradual process, like a progressive impregnation. Slowly, during the nine months of pregnancy, certain parts of the mother's astral body get more and more attached to the baby's

---

[1]*L'Agenda de Mère*, Institut de Recherches Evolutives, Paris 1981, Vol.10, 22 November 1969.

Pregnancy and Gynaecology (continued)

astral body. As long as the baby is inside the mother, she won't necessarily feel much difference, for the baby's astral body is intricately linked to hers.

At the delivery, an astral shattering takes place, not unlike that of death, but on a lower scale. Some fragments of the mother's astral body suddenly separate and go with the baby. These fragments do not become an entity in the baby, but a component of its own astral body.

Other astral bits and pieces, both from the mother and the child, are released and create a situation of 'fallout' similar to what happens after death, but less intense. Naturally, the placenta will be a prime target for these various astral fragments. This is another reason why one should make sure that all the etheric placenta is eliminated in the weeks following the delivery.

Now we can understand why after a birth, in the Indian Brahmanic tradition, a family is declared in a state of *sūtaka* or impurity. The *sūtaka* of birth is not as heavy as that of death, but still calls for ritual purifications.

The Jewish tradition also considers that a woman is in a state of impurity after giving birth, as clearly expressed in the third book of the Old Testament:

"When a woman conceives and gives birth to a male child, then she shall be ritually unclean for seven days... On the eighth day, the child's foreskin shall be circumcised. Then for thirty-three more days she shall continue in the blood of purification. She shall not touch anything holy and shall not enter the sanctuary until this purification period is complete. But if she bear a female child, then she shall be unclean two weeks, as in her impurity; and she shall continue in the blood of impurity for sixty-six days." (Leviticus xii.2-5)

At the end of this period of impurity, Leviticus then prescribes a ritual as a final cleansing. If performed by someone who knew about entities—and there is no doubt that a number of rabbis did—this ritual was equivalent to the 'checking' I recommend for every new mother one or two months after the delivery.

Another effect of the minor astral shattering which takes place after birth is to contribute to the state of emptiness, if not depression, often experienced by the mother around that time. It would be excessive to regard this shattering as the only cause of post-partum depression. A pregnancy is a tiring enterprise and one does not need to be clairvoyant to understand why the mother may feel empty after the birth. Still, the shattering appears to be an important factor in influencing the mother's emotional reactions and the subsequent psychological transformations she may go through.

The main consequence of these astral transfers is that most human beings inherit a number of undesirable tendencies while in their mother's womb. Note that this does not happen entirely by chance. When observing the mechanisms by which wandering souls choose their parents, one realises that there are usually good reasons why a particular soul is driven to a particular mother. In the astral worlds, like attracts like. A kind of resonance takes place between the dispositions of the baby-to-be and that of its mother-to-be. In other words, a soul is attracted to parents whose astral bodies have some kind of similarity to its own.

However, it is not at all rare for a soul to be attracted to a mother because of certain qualities of hers, and then unexpectedly inherit a whole package of other ones. The people who think that nothing is left to chance in the mechanisms of reincarnation overlook one basic fact—the present condition of the human astral body is a complete mess. Remember, on the astral level, you are not a person, you are a mob. You are made of a crowd of characters, and each has its own dispositions and desires. A baby may well be attracted to you because of a particular character, and then inherit features related to other ones with which it does not have any connection. A real solution to this problem can only come if parents undertake a systematic catharsis of their conscious and unconscious mind before conceiving their children.[1]

7.3 Miscarriages as purges

More than once, I have observed women who underwent psychic purges by miscarrying, even if the miscarriage occurred very early. This mechanism resembles the one we have just described. Whole parts of the mother's astral body are transferred into the embryo, and everything is expelled through the miscarriage. I have seen it happen with certain women who carried very heavy dark energies, so deeply buried or so closely mingled with their whole personality that one could not see how a release could be achieved. By heavy dark energies, I do not mean entities (which are rarely difficult to get rid of, provided a qualified clearer is at hand) but negative and destructive psychological tendencies—the darkest bits of the astral body. Then, sometimes very unexpectedly, the woman falls pregnant, aborts within a few weeks, and the dark energies disappear as if by a miracle. Each time I witness this, I can't help marvelling at how clever nature is, giving us a hand in the most blocked situations.

---

[1]See *Regression, Past Life Therapy for Here and Now Freedom,* by the same author.

Pregnancy and Gynaecology (continued)

Of course any woman who goes through a miscarriage will greatly benefit from having this mechanism explained to her. It makes it much easier to cope with the miscarriage psychologically. Moreover, a conscious perception of the purging process allows the woman to intensify it. She can willingly drop a number of inner issues in the weeks that follow the miscarriage. And if the termination has to be performed surgically, then she can prepare herself and derive the best from this difficult experience. She can push as much negativity as possible into the foetus before, during and just after the curette or intervention.

Now suppose you are a woman and you live in a society where you will have between fifteen and twenty children (like certain Catholic communities of the nineteenth and early twentieth centuries in Canada, for instance). Quite probably, you will also have a few miscarriages, some of which you won't even notice because they happen too early to produce more than a good menstrual flow. A pregnancy will start and stop after only a few days, and you won't even know it. You will just think you had a late period, or a more intense period than usual—but in reality you not only miscarried, you rid yourself of a whole astral part.

The fact that more and more women systematically use contraception suppresses the natural purge resulting from having lots of children and miscarriages. Let us be very clear: I am not suggesting that there is anything wrong with contraception, or that women should have twenty children. Having babies is not the most elegant way of purging oneself. Still it is a powerful way, and if one suppresses it by contraception, one should look at methods to replace it, by opening other avenues that allow astral releases to take place.

In the 'packs' section of Chapter 15, a herbal purification technique called 'vaginal bolus' will be described. Even though the vaginal bolus does not have even a fraction of the cleansing effect of a pregnancy or a miscarriage, it still offers an interesting purging effect. However, bolus or no bolus, the need remains for a new cleansing process to be discovered.

Whilst it is very clever of a woman to get rid of her dark astral side through miscarrying, it is not at all clever for doctors to keep dead embryos in bottles only three rooms away from their surgical theatre. From the point of view of subtle bodies and energies, modern surgery tends to multiply fatal mistakes. Embryos and placentas should never be kept in hospitals or in the vicinity of vulnerable patients, because of the obvious risk of spreading astral fragments. A preferable way of getting rid of them would be to bury them where the energy is appropriate.[1] People are completely unaware of the great harm they can do to themselves or to others when manipulating what they regard as mere pieces of flesh, but which in reality sometimes carry extremely noxious energies.

7.4 Care of the mother after a delivery

The emptiness that follows the delivery makes a woman vulnerable and at a much greater risk than usual of catching perverse energies or entities. Moreover, the astral shattering at birth and the possible complications that can be related to the etheric placenta add to the risk of being parasitised by an entity. It follows that the new mother should be as protected as possible. She should be allowed to rest and remain inside, meaning both inside a quiet room and inside herself. Draughts and cold should be strictly avoided. If the woman finds it pleasant, anointing her body with a protective massage oil can be beneficial.

The quicker the mother can recharge her energies, the safer she will be. In this postnatal time, what she needs most is warmth, both physical and emotional. A Chinese technique I found quite effective consists in using moxas in the area of the acupuncture points Conception 5 (called the 'stone gate') and Conception 6 (the 'sea of energy') located roughly one inch below the navel.

Moxas are cigars stuffed with a herb called mugwort (Artemisia vulgaris). They are used in acupuncture to warm up points softly instead of using needles. One doesn't need to be an acupuncturist to implement this very simple technique, for moxas warm up a whole area and therefore do not necessitate a precise location of the points.

Just get one of these moxas from a Chinese shop that sells herbs. Light it with a candle rather than a match, for it takes a certain time to light such a thick cigar. Then use it to warm up the area one inch below the mother's belly button. Keep the burning tip of the moxa about one inch away from the skin.

Moxas are made of compressed herb wrapped in thin paper. You have to move the paper slightly to uncover about one millimetre of the stuffed herb, to allow it to burn freely. If you don't, the moxa will not burn properly. However, do not uncover the herb too much or the moxa will flare up.

---

[1] A crossing of earth lines in a remote bush area, for instance. See Chapter 12 in *Awakening the Third Eye*.

Pregnancy and Gynaecology (continued)

The proper distance from the skin is to be adjusted according to the feedback given by the mother. The sensation should be that of a pleasant and comfortable warmth. If the mother doesn't feel enough heat, move the cigar's burning end closer to the skin. If it becomes uncomfortably hot, move it away a little. There is no benefit in burning the skin. Don't forget to tip the ashes into an ashtray from time to time, so that they do not fall off and burn the skin.

Continue for five to ten minutes. Repeat two or three times per week during the first month following the delivery, and then once or twice per week during the next two months. If the weather is hot, repeat only once or twice a week for two or three weeks after the delivery, then stop. Moxas generate heat in the body, and one should therefore be very careful with them during a hot summer: an excess of heat might be detrimental to the system.

Due to their thickness, moxas can't be extinguished like cigarette butts. You need to smother them by planting their tip in an ashtray filled with sand, or simply in the soil of a pot plant. This simple technique is usually very effective at restoring the mother's energies and correcting little problems such as excessive sweating following the delivery.

While we are on the subject of moxas, a short digression: If the new mother can't breast-feed because the milk is not flowing, use a moxa to warm up the middle of the sternum area (Conception 17). I have never seen this fail.

A remarkable technique to refill the new mother's belly energy was taught to me by a Vietnamese Taoist master called Tam Long.

Wait about one week after the delivery and then get half a dozen rocks, each approximately the size of a fist. Heat up the stones in an oven at 240°C for about forty-five minutes. Then wrap the hot stones in a blanket, and put them on the mother's belly. Wrap the mother herself in plenty of blankets, and let her enjoy the warmth for three quarters of an hour (until the rocks become lukewarm or cold). No burning sensation should be felt during this technique. If the rocks are too hot, it may be necessary to wrap them in more than one blanket to achieve a comfortable temperature.

Note that the rocks cannot be replaced by a hot water bottle. In terms of elements, it is an 'earth-warmth' that must be communicated, not a 'water-warmth'.

Oriental beds are not as sophisticated as ours. They are often made of a simple wooden frame with interlaced strings. If such a bed is available, one can implement the full technique by placing hot charcoals and ashes under it, so that heat reaches the mother's back while her abdomen is being heated by the rocks. The effect is particularly pleasant and warm, recharging the kidney area at the same time as the belly.[1]

Repeat this practice two weeks later, and once again two weeks after that. According to the Taoists, whose science of longevity is great, this technique is excellent for recharging the mother's energy and preventing disorders and diseases due to emptiness. It is also said to help the abdominal muscles and skin become firm again, and to limit the formation of flabby rolls of fat.

The same technique will prove beneficial in recharging energies after a miscarriage or an abortion. With abortions, there is no need to wait a week. First implement the technique on the same day or the day after the intervention. Then repeat once or twice in the following weeks.

For at least two or three months after giving birth to a child, the Taoists strictly caution the mother against eating oysters, snails or any type of shellfish. Cucumbers and watermelons should also be avoided during the same period. The logic behind these dietary recommendations is simple: the mother is in urgent need of warmth, and therefore cold watery foods are considered to be detrimental.

While we are discussing Chinese techniques, it may be of interest to mention the point *Zhu Bin* (Kidney 9). In acupuncture, this point is used systematically during pregnancy to correct hereditary defects, in other words to avoid some of the mother's negative features being communicated to the child. Whether this can be applied to astral transfers or not, it makes sense to have recourse to the points of *Zu Shao Yin* ('Kidney' meridian) to try to preserve the baby's integrity during pregnancy.

---

[1] A similar practice can be found in the Indian tradition, where the custom was to put hot ashes and charcoals under the new mother's bed as early as two hours after the end of the delivery. But the custom was also to leave all windows and blinds closed while the fire was burning for ten days, thereby creating a suffocating atmosphere in the room.

CHAPTER 8

HOW DOES ONE CATCH AN ENTITY?

8.1 How easily is the defence system broken through?

Not that easily! For instance, one does not catch an entity by eating at a restaurant, giving a friend a hug or wearing grandma's socks. Nor does one catch an entity by using public transport or public toilets, or even by walking through a graveyard at night. It must be repeated over and over again that it usually takes a high risk situation and an accumulation of negative factors to enable an entity to come in.

On the level of energy, a human being is naturally protected by a number of mechanisms. The etheric body has its own defensive layers. Just as the physical body has an immune system and various other ways of preserving its integrity, the etheric body has a whole range of energetic devices that constantly preserve it from the penetration of anything foreign. Apart from a few exceptional cases, an entity can only come in if these natural defences are temporarily put out of action by a trauma or another unusual circumstance.

This leads us to consider two main categories of reasons why an entity breaks in and takes up residence:
• a temporary collapse of the natural defence system;
• high risk situations where, due to particular circumstances, the danger of contamination is much greater than usual. In such cases the natural defence system may prove insufficient to prevent an invasion.

When investigating how a client caught an entity, there is usually a combination of these two factors in varying proportions. This certainly applies to the two high-risk situations described in the last chapters: after the death of a relative, and for a woman after an abortion, a miscarriage or the birth of her child. In this chapter we will look at the other most common circumstances in which an entity may break in and get attached.

8.2 In the womb

> **Case study 8.1** Forty-four year old woman, Catholic nun.
>
> *What does it want?* –It likes envy. It creates envy, for things I can't have.
> *What happens to it when you feel envy?* –It gets fat, and then I get a headache. It feeds on my frustration.
> *If it could speak through you, what would it say?* –Hate them all and get back into your shell.
> [Spinning backwards in time with the ISIS technique.]
> –I see myself in a cradle. The baby had it [the entity], it was all blocked by it. And it gave it eczema. It created frustration, like an itch, and the eczema was part of it. This thing is like an itch.
> [Spinning backwards again. At one stage, without being told to do so, the client takes a curled foetal position.]
> –In the womb. It came in the womb because my mother was upset. It's my mother's. It came from my mother, at the beginning of the pregnancy. She did not want me because it was too soon after my sister. I don't like looking at this. She was very upset and it came into me from her.

From the findings of the clients themselves, it seems that while in the womb the baby can be vulnerable to catching entities. What are the reasons for this?

In the last chapter, we saw the 'purge' process by which the mother can throw out parts of her astral body into the baby. From the point of view of the baby, this is a borderline case between catching an entity and inheriting an unwanted astral part. Apart from that, the baby in the womb is completely dependent on the nutrition coming from the mother. This should be understood not only as oxygen and nutrients, but also as emotional nurturing. The baby is totally open to the mother's emotions. Whatever comes from her is received without any kind of barrier. As adults, we can hardly imagine what this means, because of all the layers of emotional protection we have built around ourselves over the years. The baby has none of this yet, so any emotional wave coming from the mother is received a thousand times more intensely than we usually allow through our defences.

Consequently, when the pregnant woman is in peace and harmony, sending love to her baby, the latter is ecstatic, sunbathing in the love and developing the roots of its future self-confidence. In the ISIS state, clients often discover that being fully wanted and loved while in the womb is the foundation of a number of qualities later on in life, such as self-esteem and

How does one catch an entity?

assertiveness, the capacity to be sure of oneself and trust one's personal resources.

However, if the pregnant woman does not really want the baby, or if she is distressed by any other circumstances, the baby will receive the negative waves without protection. This is the equivalent of what we as adults would regard as black depression—reaching the bottom of the pit. Even though the foetus is by nature extraordinarily protected on the level of energy, this can create breaches in its defence system and allow the penetration of an entity.

All this takes place at a time when the baby's different envelopes are being generated and organised (in particular its physical and its etheric bodies). The fact that the etheric is undergoing an intense formative work and needs to draw ample energies from its environment is another factor that makes the baby more vulnerable.

Case study 8.2 Thirty-three year old woman, masseuse.
Seven years before, she was a prostitute and a heroin addict. We explored an area that felt 'darker, heavier, denser' and was located on the left side of her body, behind the ribs.

What are you feeling, now? –The presence is angry because I am talking about it. It's kind of attached to my whole left side.
Does it feel more like something foreign or like a part of yourself? –Foreign. It feels different from the rest of me.
What does it want? –It wants me to fail and take on suffering. It's concerned only with hatred and suffering, and with punishing me. Because I've done wrong things in my life, and also because I have the capacity to be very happy. It just wants total destruction, holocaust: everybody dead, annihilated, disfigured. Desperate, and dependent on conditional love. It especially wants me to suffer, because I'm not taking any more drugs and I'm no longer trying to destroy myself.
What does it look like? –Translucent. It's got tentacles, covered with gunk, like mucus. When I start talking about it, it lashes out with its tentacles and tries to render me immobile, and silent.
Have you ever seen it before? –No, never. It's the first time today.
[Spinning backwards in time with the ISIS technique] –I'm about three. It's already there. It makes me cry. And it manipulates me to do weird things, like terrorising the little boy that lives next door.
[Spinning backwards again] –I see my mother. She is pregnant with me. Early pregnancy. She is very unhappy. She is drinking and she is smoking.

> She looks so pale and unhealthy, she looks nearly dead! She is in her early thirties. I have never seen her like that before. My father hits her. He is quite violent. I see my mother and she is terrified. That's when it came in. Suddenly, one night, when he was beating her. It was in the room, waiting; and it came straight into her. Straight into her womb, into me.

Even though the entity was found to have been attached to the client before she was born, she clearly identified it as something foreign. The same applied to the client in case 8.1.

In both cases (8.1 and 8.2), the entities were explored and cleared according to the method that will be outlined in Chapters 14 and 15. From the point of view of the clearer, the fact that the entity came to be attached at such an early stage in the development of the client does not make it any more difficult to clear. As far as clients are concerned, however, it may have created major emotional problems that won't be resolved just by clearing the entity, and call for some in-depth psychological work of re-construction.

8.3 Twins

Twin births are another important situation in which a foetus can catch entities. Presently, in the human species, about 1% of births are twins. However, the number of pregnancies which start with twin embryos is significantly higher. One of the two embryos stops its growth and 'dies' in the early months of the pregnancy, and everything continues as if there were only one baby. The physical remnant of the dead embryo will be found with the placenta after the delivery.[1] The problem is that the twin embryo was not only made of flesh. This situation is quite similar to the one we described for miscarriages and terminations. The non-physical parts of the twin may remain in the womb, and possibly attach themselves to the baby as an entity.

> **Case study 8.3** Twenty-nine year old man, physician.
> He had been feeling a 'heaviness' around the gall-bladder area for years, for which no medical explanation could ever be found.

---

[1] If the twin is aborted in very early stages of the pregnancy, it may be extremely difficult to find any remnant of it in the placenta, which does not make it easy to establish proper statistics.

How does one catch an entity?

> –It looks like a foetus, about five or six inches long. It's made of a kind of blue-grey translucent jelly, exactly like a foetus. It's attached to my right side, under the ribs, just where I get this uncomfortable feeling all the time.
> *What is the connection between the foetus and the uncomfortable feeling?*
> –The foetus is taking all the bloody space in the area. It's a half-foot long foetus sleeping in my liver! No wonder I feel uneasy.
> *What is it doing?* –Nothing. Just sleeping. Sometimes it moves around, but it remains attached to my right side by a kind of cord. Sometimes it gets into my head, and I get these terrible headaches. The only thing I can do to stop them is go to bed and fall asleep.
> *Is there any emotion or feeling related to it?* –It's angry. Angry at me. Like a kind of grudge.
> [Spinning backwards with the ISIS technique] –It's dark. A dark space. In the womb. It was there before me. It came first, I arrived second. We are both the same size, and in the same foetal position. But we are smaller than he is now. He is on my right. He is attached to my right side, like now. I can feel his strong presence on my right. He is a powerful person, a fighter. Then something goes wrong for him. He starts fading. He disappears. But his [etheric] foetus body stays close to me. It was not as big as it is now. It kept on growing a bit with time. But empty, just a shadow of the presence there was before. The anger is because I have taken its place.

In some cases I have seen clients describe a fight between them and the other twin. The fight does not happen physically of course, but on the astral level.

> **Case study 8.4** Thirty-five year old man, artist, who was feeling a dark presence attached to his shoulders.
>
> [Spinning back into the womb] –I see another person coming into the cave. Another soul. I don't want him to come in. It's like now I have to share the space with him. But I seem to be growing faster than him. He is struggling, doing anything he can to stay in. He is even pushing me, kicking me. He is trying to kick me out. But he's lost. I'm growing and he is shrinking. So he tries to push me out and take my body, and I have to hold fast. That body is mine, I'm not going to let him in. He can't take my place inside the body, so he sticks to my back.

In my practice, I have come across a number of clients who discovered similar entities which originated from a lost twin (and who, needless to say, had never thought of that possibility before undergoing ISIS). This leads me to consider that the proportion of twins which are aborted in the early stages of pregnancy could in reality be higher than physiologists presently believe.

8.4 Drugs and surgery

During sleep, the upper complex (astral body + Ego) moves away from the lower one (physical + etheric bodies) and directs its activities towards the astral worlds. While this happens, the lower complex (physical + etheric bodies) is certainly not left unprotected. However, its defences are not inexhaustible either, and breaches may be created due to a number of reasons.

Among the most common factors is the consumption of certain drugs, either for intoxication or for medicinal purposes. Alcohol, sleeping pills, pain killers, anaesthetics and the whole range of narcotic drugs from marijuana to heroin, all create a disorganisation of the defence system and can cause breaches.

Of course, they do not all have the same disorganising power. For instance the probability of catching an entity due to one glass of wine or one cigarette of marijuana is virtually nil. On the other hand, most surgical anaesthetics leave you completely vulnerable and open to the penetration of an entity.

Case study 8.5 Fifty-two year old man, school teacher. Exploring an entity located in the lower abdomen.

What does it look like? –It looks like a black panther, slinky, with bright brown eyes. The panther is not facing me, his tail is towards me.
Does it enjoy certain foods? –Meat. He always makes me eat more meat than I need.
Are there any emotions or feelings associated with it? –He makes me feel shy. He feeds off my power. He eats my energy and my confidence. He doesn't want me to express my full potential.
[The client goes through the spinning mode of the ISIS technique] –I can see it coming in when I was four. It came while I was being operated on for a hernia. And after the operation I had a complete change of personality. I was a happy-go-lucky child before the operation. After the operation I became very shy. The panther made me lose my confidence. I started shrinking back.

How does one catch an entity?

As a medical practitioner, I had the occasion to observe first-hand how easy it can be for patients to catch entities during surgical operations. All the required circumstances are gathered. Firstly, anaesthesia forces patients out of their body and stupefies their defence system. Secondly, the skin opening made by the scalpel is accompanied by a breach in the defensive layers of the etheric body, creating an ideal gateway for the penetration of an entity. Thirdly, surgical theatres are built in a way that tends to accumulate perverse energies and entities.

When I started my medical studies, I saw the last of the friendly surgical theatres with open windows. Soon after, they were replaced by air-conditioned closed cells, where sunlight never reaches. Any room that never receives sunlight and in which the only way to change the air is through activating an air-conditioning system is a high risk place for entities: it attracts them and provides an ideal space for them to stay undisturbed. Whenever an entity happens to come in, it is really hard for it to find a way out, apart from sticking to a patient or a member of the staff.

Of course, I am *not* suggesting that all people who undergo surgery end up with an entity. Most patients come out of the intervention energetically intact. The point I want to make is that, if many people happen to catch an entity or a perverse energy during an operation, it is for reasons that could be avoided most of the time. One could certainly limit the problems of energy contamination by implementing a few simple measures, such as using local rather than general anaesthesia whenever possible, carefully selecting the place where theatres are built, and designing them with many windows.[1] It would be quite possible to leave the windows open when the theatre is unused, and seal them to sterilise the atmosphere prior to an intervention.

Of course, a much higher degree of protection could be achieved by including someone who knows about energies and entities in each surgical team. Thus a clearing could be undertaken each time something suspect is perceived in the theatre.

8.5 Emotional and physical shocks

Traumas of various kinds, either emotional or physical, can create a temporary collapse of the etheric-astral system of defence. Remember the Chinese way of expressing a great emotional turmoil: "My *Po* are all fidgeting and my *Hun* are totally confused!" The highly disorganised inner state that follows certain traumas can be favourable to the penetration of an

---

[1] Built on an energy well, a surgical theatre could work wonders. See Chapter 12 on earth lines in *Awakening the Third Eye*, by the same author.

entity. For instance, remember case study 1.5 (the little red ape that liked coffee very much). It was during the panic that followed the news of her father's death that the client was invaded by the entity.

Case study 8.6 Forty-four year old man, working in a bakery.

What does it look like? –It's a dark cloud. It's to do with the heart. It's like something pressing on my chest, like something choking me or strangling me. It hates me, it wants to hang around and drain me.
To drain you? –It takes energy from my heart by making me sad. When I get sad, that feeds it, and then I'm all empty. I taste all the cakes and I stuff myself with hot croissants and pastries.
What sort of food does the cloud like? [Long silence, then] –Sugar. I think it likes it when I have sugar.
What does it look like? –It's like a dark blob, like a dark patch in my energy around the heart. But it's not my personality, it's not me; it's from outside. It's inside me and it's not inside me, all at the same time. It's with me all the time, but it's stronger when I'm alone. It started when I was living in America, four years ago, close to a forest. I had a big fight with my girlfriend. I was all trembling. Then I went out into the forest, it was night. And something weird happened. I couldn't say what. But the day after, my whole back went crazy, I couldn't move my left side. It had never happened to me before. That's when I noticed the presence. And then for months I would wake up in the middle of the night and I'd feel the presence around me. The same presence as in the cloud. Sometimes I was so scared I could not even scream for help, I'd remain motionless in bed like a stone.

Similarly, all huge emotions, such as panic, rage, desperation, can be responsible for a collapse of the defence system. Physical shocks are also among the common causes that may allow an entity to penetrate.

Case study 8.7 Twenty-seven year old woman, secretary. Exploring a 'darker and denser area' in her energy with the ISIS technique.

What does it want? –It wants to get out. It's very angry, total rage. It's angry because it's stuck. It just wants to be free. So it screams and shouts. It's mad because I can see it.
[Spinning backwards in time.] –It came when I was little; one, or one and a half. I burnt my hand on a stove. I had a shock and I left my body. That's

How does one catch an entity?

when it came in. It feels like a dark presence that won't let me live my life. It wants my body. It wants to live there.
Are there foods that it enjoys? –It likes meat. And thick foods—cakes, ice-cream and all that stuff. It says "You can't do without me. You wouldn't know what to do without me. I've got you, you can't stop, you can't get rid of me. I'm in you and I'll stay there. I've got you."
What would happen to it if you died? –It would die too. Or it would have to find someone else.

8.6 Inviting an entity in

Case study 8.8 Fifty year old woman, housewife. Exploring a 'darker, denser, heavier area' perceived on the right side of her body inside the ribs.

What does 'the thing' look like? –It looks like a dark blob. A kind of amoeba. About three inches wide.
What does it want? –Drain me. Take my life force.
What does it gain out of it? –I'm its survival means. I make it feel alive and stay alive. It gives me headaches. It takes all my courage away. It makes me become grey, like it.
Does it sometimes create voices in your head? –No, but it used to. When I was a child. It was keeping me company. I was terribly lonely as a child. I was always praying for someone to come and be with me and play with me. I called it for days. One night I remember I went out in the garden at night and I knew it was wrong. I was about seven or eight. Nothing special happened that night, but I knew I had done something that wasn't right. And after that there was this presence with me. It was keeping me company, reassuring me when I was afraid. It took care of me if I cried. My mother was never at home, she was working.

8.7 Children asking an entity in

I have come across several clients who, like this one, recalled having willingly attracted their entity during their childhood. The pattern is more or less the same in each case: the child is lonely, insecure, afraid, or just wants to have fun. After calling for days, or weeks, an entity arrives. Remember what we discussed about the fragments: in the astral worlds, vibrations can be felt from afar, and attract corresponding forces. If you drink lots of beer, you naturally tend to attract beer-drinker fragments. If you are desperate for company or protection, you tend to attract nurturing ones. This is not inevitable, of course, but it does happen sometimes.

Usually, if a fragment is attracted, children will be protected by their natural defences. In some cases, however, the child pushes the game a bit too far, or some kind of accident takes place—and the entity is in!

In similar cases I have observed, clients related that at first they were quite happy to have the presence with them. It kept them company. It sometimes even played with them. It gave them the impression of being looked after. It was only many years later that they came to realise that 'the thing' also acted as a parasite, draining their life force and creating the secondary symptoms we described in Chapters 1 and 2. Often they had completely forgotten about the presence they felt as a child, and it was only when undergoing the ISIS process that they established the link between their current problems and the 'companion' of their early years.

In this section, let me also briefly mention that people who are involved in magic, spells, etc. can easily put themselves in the situation of 'inviting an entity in', whether they realise it or not. We will see a spectacular example of this kind in the chapter on 'extraordinary entities' (case 13.3).

8.8 Other factors

Let us look at a few other traumatic circumstances when a collapse of the natural defence system may take place and possibly open the way for an entity to come in:

• With any drug, past a certain level of intoxication.

• When very sick, in particular when there is haemorrhage or intense loss of fluid, or when very weak, e.g. with chronic illnesses which create a major depletion of energy.

• Certain 'hard' medical treatments, apart from surgery, can also allow an entity in. Electroshock therapy, for instance, appears to be a high risk situation for catching entities.

• Natural disasters, such as earthquakes, or man-made ones, such as battles, war, fires, etc. Moreover, from what we have described about the shattering of the astral body after death, we can see that any catastrophe that kills tens or hundreds of thousands of people will be followed by the release of a tidal wave of astral fragments.[1]

---

[1]In 1991, at the conclusion of the Gulf War, when a large number of Iraqis were killed, I witnessed in the astral the release of a phenomenal amount of fragments. Moreover, it appeared to me that the clouds of dark

How does one catch an entity?

Some other factors can contribute to a slow exhaustion of the etheric protection layers, thereby allowing an entity in.

• Sleeping on noxious earth lines (sometimes also called earth-ray lines or ley lines). Living in a house full of noxious lines often leads to catching multiple perverse energies and possibly entities.[1] Spending long hours of meditation on a noxious earth line can also result in catching an entity. For, during meditation, a great opening of one's energies takes place. One should therefore carefully select the place where one meditates.

• Sleeping or meditating on top of an underground creek or sewage pipe, or on top of a tank of stagnant water, can be quite destructive to your etheric body and open the way to perverse energies and entities. A mass of stagnant water (as in a water bed, for instance) can attract several 'unattached' entities.

• Exhausting the body by not sleeping enough, working during the night and sleeping during the day, working beyond one's limits over a long period of time, etc.

8.9 Can one catch entities during sexual intercourse?

Most of the time, no. However, making love involves not only an exchange of emotions and body fluids, but also of energies. Two types of things can happen during intercourse. Perverse energies can be transferred from one partner to the other. Another possibility is that an entity that is floating around takes advantage of the situation and invades one of the partners. The probability of an entity jumping from one partner to the other during intercourse seems very slight to me, simply because once an entity is nested in a person it doesn't move out easily. One has to implement a very special technique to expel it. Therefore I don't see that one could get rid of an entity by sending it into one's partner during sexual intercourse.

The situation may be different for people who have multiple partners and a raging sex life. In that case, an exhaustion of essential energies may

---

smoke created by the massive burning of the oil wells had an unexpected effect: it created a thick and confused energy, which densified the fragments and kept a large number of them trapped in the area. This was responsible for a depression in the planetary balance of energies that lasted for months. It also created a great deal of entity-related problems, not only in the area, but also in various parts of the world. In particular, it was my perception that the cholera epidemics that took place in South America in the following months were directly related to this disruption of energies and to the release of fragments.

[1]See Chapter 12 in *Awakening the Third Eye*.

favour the penetration of one or even many entities. (Due to their cravings, fragments are often attracted to people who have a particularly intense sexual life.) However, under normal circumstances, there is no need to be overanxious about catching an entity through sex. Again, one does not catch entities that easily!

The situation is different with perverse energies, i.e. bits of noxious etheric matter which, unlike entities, do not have any astral consciousness attached to them. Entities do not come in easily. Once they are inside, they are extremely stubborn and usually only leave if a proper clearing is implemented. On the other hand, many perverse energies can penetrate or leave much more easily. Thus it is quite common for masseurs to catch perverse energies while massaging their clients (and not uncommon for clients to catch perverse energies while being massaged). If perverse energies can be exchanged through massage, it is obviously even more possible to catch one during the sexual act. During and after orgasm, and at other stages of intercourse, the body of energy opens significantly. This contributes to the sexual experience, but also makes one more vulnerable to perverse energies.

8.10 Sexual intercourse and Taoist days

It is easily noticeable that, on certain days, sexual intercourse seems to be more tiring than on others. It leaves the partners (or one of the partners) with a feeling of emptiness and fatigue. The Taoists, who were experts in the art of longevity, observed this fact and related it to their understanding of energies. According to their system, the quality of energies varies with the Moon phases and the cycle of the seasons. Their conclusion was that having sex at particular times in these cycles can create a depletion of essential energies and leave the door open for the penetration of perverse energies, and even entities.

One critical time is the Full Moon. From a Taoist point of view, sex twenty-four hours before and after the Full Moon is considered potentially harmful, even if there is no ejaculation or orgasm.

Another critical time is the New Moon. The restrictions are even tougher if the New or Full Moon is accompanied by an eclipse.[1]

It must be stressed that it is not only the ancient Chinese who considered sex at the New and Full Moon to be dangerous and therefore to be avoided, but also Tibetan and Indian masters. In particular, one finds clear references in Sanskrit texts to the possibility of catching an entity if having

---

[1] Eclipses can take place only at the New Moon or the Full Moon.

How does one catch an entity?

intercourse on these days, and even more if there is an eclipse.[1] The Tibetans consider the Moon rays at the Full Moon as potentially dangerous in relation to energies and entities. According to them, one should not sleep outdoors during the night of the Full Moon. Moreover one should block the moonlight with blinds, which applies even if not having sex.

Interestingly, from the Taoist perspective, it is not only sex that is forbidden at the Full Moon, but most therapeutic acts. According to the Taoists, one should never insert an acupuncture needle on the day of the Full Moon, or have a surgical operation or a dental treatment. For at the Full Moon the etheric body is more open and externalised than on any other day of the lunar cycle. Therefore any breach may create a major energy leakage—hence the restrictions regarding treatments as well as sexual intercourse.[2]

The other main times which are deemed inappropriate for sex by the Taoists are the solstices and the day before, during and after the equinoxes.

Another period that both the Chinese and the Hindus consider dangerous for sexual intercourse is menstruation. Similar restrictions can actually be found in nearly all cultures, from the Old Testament (Leviticus xviii.19) to Australian Aborigines. From the point of view of subtle bodies, when women menstruate they get rid of negative etheric energies. This has many benefits for their own health, but it is usually considered that if intercourse happens during this period, unfavourable consequences may result for both partners. For the woman, it may disturb the natural course of the menstruation. For the man, there is a risk of catching negative energies.

Are these restrictions to be taken seriously by twenty-first century people? The approach of the Clairvision School is always to pay more attention to direct experience than to theory. My purpose is therefore to draw the reader's attention to facts that may seem unusual, but have been emphasised by a great number of authorities on energy in different civilisations. It is up to you to assess them in the light of your own experience.

---

[1] See for instance *Caraka-Saṃhitā*, *Nidānasthāna* VII.14.

[2] However, one should never forget that the Buddha is said to have been conceived at the Full Moon, and born at the Full Moon. All the major events in his life are said to have taken place at the Full Moon: his renouncement, his enlightenment, the beginning of his teaching, and finally leaving his body. Thus if the Buddha's parents had strictly followed the above mentioned restrictions, there would never have been a Buddha!

CHAPTER 9

ENTITIES OF VARIOUS KINDS

9.1 Fragments

In this chapter, we will go through a brief inventory of the main types of entities that can get attached to human beings.

First of all, we examined the fragments, meaning the astral bits that break off from the shattered astral bodies of the dead. Then we studied a few particular types of fragments, those left after a miscarriage, a termination or a delivery, or after the natural 'death' of a twin in the womb.

Fragments are not the only type of entities, but they are by far the most common. However, they can sometimes be confusing because even though they come from human beings, they do not necessarily look human. We have seen some which assumed a semi-human shape. In the case studies, however, we have also seen entities that appeared to clients as blobs, clouds, amoeba, octopi and various other monstrous shapes. Most of these were fragments too. Another confusing point is that fragments often appear much worse than the people they have come from. I have seen a number of cases where there really could not be much doubt: the entity had appeared in the weeks following the death of a close relative; its presence felt similar to that of the person while he/she was still alive; and it presented many common psychological features that left no doubt as to its source. However, the client would comment: "I understand that this is a fragment from my deceased wife, but still I can't understand why it's so ferocious and noxious. She was not like that!"

It must therefore be emphasised that fragments are just fragments, not people. Fragments are characters that have broken off from the astral body of a human being. When they were attached to the astral body, they were under the influence of the rest of the personality, and repressed by a number of factors. They never appeared exactly as they were, because they were lost in the middle of a mob of other characters with other inclinations and dispositions.

Entities of Various Kinds

Suppose a character was a passionate chocolate eater. While alive, the person probably indulged in it from time to time, but not from morning to night. Our society imposes certain rules and limitations, and people cannot usually obey the impulses of their astral body without restriction. Once dead, however, none of these limitations apply any more. The fragment is totally de-repressed. It becomes a chocolate devourer wandering in the space, and screaming for its cravings without inhibition. Likewise, other fragments with different emotions or desires suddenly appear as they really are. Often, that is not very pleasant. For the same reasons, many fragments will not bear much physical resemblance to the person from whom they have broken off, and will appear in monstrous forms.

9.2 Ghosts?

At this point let us tackle a basic question: what is a ghost? Think of the traditional cliche of the ghost that lives in a castle. The ghost appears every night at exactly the same hour. It executes exactly the same actions every night, in exactly the same order. This sounds very much like a fragment which endlessly repeats what has been imprinted in its substance. From our perspective of subtle bodies, a 'ghost' appears to be an extremely crystallised astral fragment, possibly coated with etheric layers, and bound to a house instead of being attached to a person.

If a person was born in a castle or a mansion, lived a whole life and died in it, and was buried somewhere in its cellar, it would not be surprising if fragments stuck around afterwards. As we have seen, repetition is an essential factor in creating deep imprints in the substance of the astral body. The fact of living one's whole life in the same place, especially if one loves or hates the place, contributes to the formation of extremely crystallised fragments.

I will only briefly mention the poltergeist effect that may accompany some of the 'ghost' cases, because this remains quite rare among the relatively large number of entity cases. (By dwelling too much on a few spectacular cases, one tends to create false ideas and cliches that divert attention from the reality of the phenomenon.) In terms of our system of subtle bodies, how can one explain that certain ghost manifestations not only take place on the level of etheric energy and astral consciousness, but also involve certain physical phenomena, such as physical sounds or movements of objects? The answer lies in the fact that the transition between the physical and etheric layers is gradual.

Let us use the example of the musical scale. The first octave corresponds to the physical world, the second to the etheric world. At the end of the first octave and the beginning of the second one, we find a few layers that are half way between physical and etheric. They are the most refined of the physical layers, and the densest of the etheric ones. These are the layers in which the poltergeist type of phenomena take place. As above, so below. Similar transitional layers can be found between the physical and etheric bodies. The densest layers of our etheric body are nearly 'physical'. Suppose a fragment has reached a very high level of crystallisation. It may well be able to retain some of these densest etheric substances around itself.

Thus, in terms of subtle bodies, a ghost that creates poltergeist manifestations is an extremely crystallised fragment surrounded by the densest etheric energies. It is through these layers on the borderline between etheric and physical that fragments can sometimes have an action on physical matter and produce poltergeist phenomena.

9.3 Perverse energies and entities

Apart from fragments, what else can act as an entity and parasitise human beings? In Chapter 3, we discussed the concept of *Xie Qi*, or perverse energies. In terms of subtle bodies, a perverse energy is a bit of etheric substance that has infiltrated the human etheric body and is detrimental to its functioning. Strictly speaking, a perverse energy is nothing more than etheric substance, whereas an entity is made of astral substance, with or without etheric substance around it. Due to its astral part, the entity has a certain mental consciousness of its own with thoughts and emotions, while the perverse energy doesn't have the same level of organisation.

In practice, however, the borderline between perverse energies and entities is not necessarily so clearly defined. It is not rare for certain perverse energies to get linked to chips or diluted clouds of astral matter. This endows the perverse energy with a faint and blurry presence. It may also animate it with elementary emotional waves.

When exploring such a 'thing', the client cannot usually feel or see much. Of course, one should always first suspect that it is a full entity hiding behind a nebulous cloud and insist, go deeper into the ISIS state, ask questions, etc. But as the Chinese say, it is very difficult to find a black cat in a dark room, especially when it is not there. Sometimes one has to accept that the 'thing' is nothing much more than a perverse energy. When I find such a borderline perverse energy in a client, I tend to process it in the same way as an entity, simply because it works remarkably well as far as the main purpose is concerned, that is, getting rid of it.

Entities of Various Kinds

9.4 Elementals and nature spirits

Elementals are non-physical little beings that stand behind earth, water, air or fire, or behind flowers, trees or other plants. They can be related to the various spirits of nature, echoes of which can be found in nearly all the mythologies and folklores of the planet.

Case study 9.1 Forty-four year old woman, shopkeeper. She felt there was 'something wrong' with her left shoulder: no pain, but a more or less constant feeling of heaviness, as if her left shoulder could not move as easily as her right one. In the ISIS state, she immediately identified a 'darker, denser, heavier patch' perceived in her back close to her left shoulder blade.

What are you feeling? –It's like a draining, exhausting type of feeling.
What does it look like? –On the left side of the spine, there is a darker patch. The spot looks like a dark brownish walnut, with a rough surface.
What does it want? –It wants to keep me tied down, so it's got a place to be. It drains my energy. It makes me feel lethargic.
Are there some foods that it enjoys? –Sweets. Starchy foods. And it makes me eat too much bread. It does not like the cold, it likes the Sun. It doesn't like it when I go out. It makes me withdraw from people and hide, stay at home and do nothing, choked up with grief and my inability to communicate.
What's inside the walnut? –I can see a lot of miniature people. They are dressed like in biblical times, with long dresses and turbans. They are not doing anything special. They are just standing there quietly.
If the walnut was to leave you, where would it go? –Deep into the earth.

Indian, Chinese and Tibetan records on entities all mention the possibility of elemental spirits of nature behaving like entities at times, and creating various health or mental disorders. If we analyse the symptoms reported by this client, we can see that they all correspond to qualities that ancient medical lores (both eastern and western) used to relate to the earth element: heaviness, lethargy, inertia, feeling cold and craving for the Sun's warmth, etc. The 'little people' are one of the ways elemental beings can be perceived. The elemental beings behind the elements have the reputation of being hard to see, for they are tricky and like to hide.

As is often the case, during the clearing the client could see the walnut moving upwards and disappearing out of her body. The uneasy feeling

Entities

in the left shoulder vanished in the days that followed the clearing and did not come back.

While on the topic of elementals and nature spirits, it may be of interest to mention that in Australia we have some particularly strong sacred sites and land energies, with which the aboriginal culture was well acquainted. One does not even need to drive for a week and reach the desert to find them. They are everywhere. The whole of the Sydney area, for instance, is full of them. Some of these sites have a wonderfully subtle and healing energy. Others are completely incompatible with human life. Aborigines respected them for their overwhelming strength, but declared them unfit for human habitation.

Unfortunately, Australian architects and developers have not taken this geography of energy into account. A number of dwellings have been built on sites where human beings should normally never stay. I have seen such houses in which the residents and their pets caught entity after entity, and many perverse energies. The spirits of the place were constantly waging war on them, and the results were sometimes tragic. Animals and people became depressed, mentally disturbed and severely ill. Some people even committed suicide. Trying to 'clear' such a house would not make much sense to me. Different places have different energies, and it is up to us to inhabit those which are happy to receive us and supportive of our vitality.

Some places may be life supportive by nature, but temporarily parasitised by a fragment or some other entity. In that case, a clearing makes sense. If, however, you live in a place that is intrinsically unfit for human life, then it is better to move than start a war against the spirits of the land. One reaches balance and mastery by harmonising oneself with universal energies, including land energies, and not by fighting them all the time.

Case study 9.2 Thirty year old woman, osteopath. Exploring a dark cloud she could feel in her liver.

When can you first remember feeling it? –It was night and we had been walking in the bush for hours. We were completely lost. There was no moon and we did not have a clue where we were. So we decided to put up the tent and spend the night there.

–All night I had nightmares of a huge black cloud coming down over me. I could feel it coming inside me, and withdraw, and come inside again. I

> didn't know what it was. It just felt very very old, like an ancient awesome presence.
> —When we woke up in the morning, we realised that we had been sleeping at the foot of massive rocks, abrupt cliffs more than one hundred meters high. I couldn't feel the awesome presence any more, but I had this black cloud in my liver. It has been there ever since.

This cloud was cleared with no more difficulty than any other entity. Nevertheless, the whole episode could have been avoided if the client had been able to use her sensitivity and feel that this particular spot was not appropriate for camping. Just by walking another hundred meters and finding a more welcoming place, none of this would have happened.

9.5 Other entities

Entity is a very general term. I have used it in this book to refer to the presences that can parasitise human beings. However, in its broadest sense the word entity can be applied to any being, physical or non-physical. So you are an entity, I am an entity, the cat is an entity. Elephants, dolphins, elementals, nature spirits and angels can all be called 'entities'.

Among the whole range of non-physical beings that constitute the creation, which can become parasites of human beings? Not all, of course, but many of them. Let us take a quick look at some of the beings described in Hinduism. Ayurveda, or traditional Indian medicine, makes a clear distinction between the cases of insanity due to internal factors, the same with any other disease, and those brought on by the pernicious influence of certain non-physical beings.[1]

In some cases, non-physical beings are said to invade people and become parasites, just as in the entity pattern described in this book. Thus, there is a particular class of beings called *piśācas* in Sanskrit. The *piśācas* are described as depraved and vicious beings. They are said to attach themselves to human beings by jumping onto their victim's shoulders and riding them like horses.

Some other entities are of too high a calibre to bother invading human beings. The *rākṣasas*, for instance, who live for hundreds of years, are fierce creatures with extraordinarily intense desires, passions and instincts. They render human beings insane just by letting them smell the foul odour of their body.

---

[1] See *Caraka-Saṃhitā, Nidhānasthāna*, Chapter 7.

Entities

However, it is not only depraved little entities or monstrous big ones that can cause insanity in human beings. If insulted or offended, the gods themselves, so respected and cherished by the Hindus, can strike even their own worshippers. They create insanity, epilepsy or other afflictions, just by looking at their victim. The *gandharvas*, a category of semi-divine beings devoted to music and the arts, can create insanity just by touching human beings; and the *pitṛs*, or souls of the ancestors, just by appearing to them. Even the *ṛṣis*, the enlightened seer-sages that dwell in the higher spheres of creation, can create madness in a human being by sending down a curse.

What is the relevance of these mythologies today? As we have seen, in the vast majority of cases, entities are quite simple little things. If one doesn't know how to deal with them, they can certainly create various troubles or illnesses. Nevertheless, if a qualified clearer is at hand, they are not much of a problem: it takes much less time and effort to have an entity cleared than a tooth fixed by a dentist. Moreover, it is completely painless. Usually, entities are nothing more than a fragment or an elemental that has gone astray. But not always! In Chapter 13, 'Extraordinary Entities', we will look at a few case studies that don't fit with this simple pattern, and where one might have to talk about dark forces or take a fresh look at old mythological stories.

Here is a case study which does not present anything spectacular, yet involves a being of a different rank from fragments or simple nature spirits.

Case study 9.3 Twenty-six year old man, taxi driver, ex-heroin addict. He came to consult me a few weeks after having a car accident. He was heavily depressed. In the first part of the ISIS session, he identified a black cloud in the area of his physical heart.

Is there any emotion associated with 'the thing'? –Fear. Fear of death. Fear that it is going to kill me.
What does it want? –It just wants to live. It lives in people's minds like a headache. If you need a headache, it will come to you and give you one. It knows exactly how to manipulate you to make you sick, and it takes a lot of pleasure in it.
What does it look like? –Like a mischievous child. Around it, my body feels very heavy. And black, pitch black. I think that I might have invited it. I wanted to hurt myself. Now I can remember that it happened when I was 12 years old. After that it was delighted when I was into heroin. This thing feels evil. It wants you to suffer before you die. It wants to harm you.

> When I used to take drugs it was very strong. It enjoyed it. It made it feel full. When I stopped taking drugs, it disappeared, but it was not gone, just sleeping inside. The car accident woke it up. It needed the car accident to be able to start again. It might even have triggered the accident.
> *If you died, what would it do?* –Go to somebody else. It would just hang around and go to somebody else. It moves around from person to person.
> *Is there any connection between this thing and your depression?* –The depression comes from it. It secretes the depression like a dark cloud into my heart.

The clearing of this entity did not present any difficulties, but I could see that what I was removing was different from the blobbish fragments usually found. This was a sharp and malevolent little being, which would have gone straight into someone else if it had not been 'recycled' by the Great Light with which entities are cleared.

This is a typical case where conventional forms of medicine are completely inadequate. Starting a treatment with antidepressants, for instance, would not have solved the problem in any way. The depression would have dragged on and had the emotional symptoms been suppressed, the little being could well have made its action felt in an even more destructive way.

Immediately after the clearing, the client could see his heart looking "clear, like pink flesh, with pink blood." The depression vanished within two days.

9.6 What entities are not

At this stage, it is important to dwell on what entities are not. Entities are not full human Spirits who have lost their way and become attached to a living human being. If four weeks after the death of your aunt you happen to catch an entity that feels like her, it is just a fragment of her astral body, not her immortal soul trapped inside you.

There are several reasons to support this statement. We have already seen a few of them. Firstly, as we have seen, entities often appear much worse than the people they come from. For instance your aunt was a most delightful old lady and, after exploration, it appears that the entity cares for nothing but sugar, sex and violence. Even though the entity's presence clearly feels like your aunt, the picture obviously does not fit. Note that this may not always be obvious at first, because entities often try to camouflage themselves behind a shroud of good intentions, if not holiness. However, one does not have to insist very much to reveal their basic instinctive motivation.

Many Chinese stories of *Kuei* convey a similar message. For example, a man is looking after a dear friend of his who is dying. The friend passes away. In the following days, the man is haunted by a spectre of his dead friend. Instead of showing gratitude, the spectre comes to frighten him, or even attack him. A sage is then consulted and explains that the spectre is not the friend's immortal soul but only his *Kuei*, meaning a fragment. That is why the apparition was so aggressive and frightening. It had nothing to do with the friend himself, i.e. his *Shen*, which has departed and is on its way to the spiritual worlds.

Another fact we have observed is that entities are usually 'monofocussed', polarised in one narrow direction, always wanting the same things or repeating the same emotions. This also indicates that they are not the whole of the dead person, but only a fragment corresponding to one particular character of theirs.

From the point of view of the clearer, it should be emphasised that nearly all these entities are only insignificant little 'things'. While clearing them, one can see them clearly. In no way could these little chunks of astral matter be taken for full human Spirits. The light and the magnitude of the Higher Self are infinitely greater, beyond any comparison with the vibrations and the low level organisation of these entities. Moreover, both from the point of view of Christian esotericism and Hinduism, it would be a metaphysical absurdity to consider that an eternal human Spirit can remain trapped in the body of a relative, or even in a castle.

The word spirit can be misleading, for it can be used in two completely different ways. With a capital 'S', Spirit means the Higher Ego or Self, the immortal part of a human being. It is the transcendental part, forever united with the Divine. With a lower case 's', however, the word spirit refers to elemental beings and various other classes of non-physical beings. Since fragments behave as semi-autonomous etheric-astral 'creatures', if one wishes one can also call them spirits.

The Spirit is the immortal flame, the part in which metaphysical freedom already shines. One does not become enlightened by enlightening one's Spirit, but by unveiling It. For the Spirit is already fully enlightened and connected with the Divine. To one who knows the Spirit, or Self, the idea that It could be trapped in the material world after death sounds more than fanciful, for if there is one part in human beings that can never be lost, it is precisely the Spirit.

CHAPTER 10

ENTITIES AND PAST LIVES

10.1 The bodiless head

Case study 10.1 Twenty-seven year old man, librarian. Exploring an entity that first appeared to him in an ISIS session as a dark cloud located in his left side, behind the ribs.

What does 'the thing' look like? –I can see a head. It looks Chinese, with a long moustache.
Is there a body that goes with this head? –No, just a head. A grotesque head, like a caricature.
Have you ever seen it before? –No. But it feels familiar.
Are there any emotions or feelings associated with the head? –Hatred. It hates me. Its eyes are looking at me as if it wanted to kill me. It says "I'll kill you. It's just a question of time. I'll attract some men who will come and beat you up. And I'll make you fall sick and you'll die."
[The client starts spinning backwards in time with the ISIS technique.[1] He remains silent for a while, immersed in the inner space, then:]
–I can see horses. Fear. Some men have come to take me. My hands are tied behind my back and they are taking me somewhere. They are brutal, I'm afraid they're going to kill me. They take me to a man in a tent. THIS MAN HAS THE SAME HEAD, but with a full body, like a real person. His features look normal, whereas the head is grotesque, nearly inhuman. The man hits me in the face. He's captured me to put pressure on my father. Somehow my father seems to be someone important, and they're try-

---

[1]The 'spinning' mentioned in many cases does not refer to any physical movement, but to a particular technique implemented in the space of consciousness during ISIS. For more details, see the practical section of *Regression, Past Life Therapy for Here and Now Freedom*, by the same author.

ing to ransom him, or maybe force him to take a certain political decision, or something like that.

[Spinning again:] —The brigand's plan went wrong. My father's men came and set me free. The brigand is beaten. And tortured. And mutilated. They take all his property, and they send him away. He becomes an outcast, a wanderer wearing rags and begging for his food. He keeps on thinking about me, hating me. He can't forget. It becomes an obsession. He prays he'll have an occasion to take his revenge. Up to his very last day, he was obsessed with hatred.

What happened after he died? —The head was waiting for me.

Where was it? —Somewhere in space, in a dark area. Waiting for me to come back. Not long after I was born it found me. It was very happy when it found me. It said "Now, I'm going to give you a real hard time". When I was a child, it used to make me feel panic in the dark for no reason. It was the same head, it was just that I couldn't see it.

In a number of cases, while exploring entities with the ISIS technique, clients re-experience what they consider to be a past-life episode, in which seems to lie the key to their present problem. If we apply our pattern of understanding to the case study given above, we find that an angry and revengeful fragment was released and waited patiently in the space before it could find its victim again. As we have seen, the fragment appears to the client as a grotesque caricature of the Chinese brigand's face.

One can easily conceive that the accumulated hatred and bitterness of the brigand were favourable conditions to generate a very crystallised astral fragment. If a man broods over revenge every single day of his life, the corresponding thought form in his astral body gets solidified and sharp. He is therefore likely to release quite a coherent and noxious fragment when he dies. Why didn't the fragment go straight to the young man in China, why did it have to wait for his reincarnation in Australia? Maybe the young man died before the beggar. Maybe he was still alive at the beggar's death, but the head could not find him then. Here is another similar case study.

Entities and Past Lives

10.2 The witch that never lets go

Case study 10.2 Thirty-four year old woman, counsellor. While exploring a fear of hurting her own children, an entity was located in the left iliac area.

What does 'the thing' look like? —It looks like a witch. The laughter and the witch face come to me each time I tune into the presence. She looks grim, and very intelligent. She wants revenge. She holds a grudge against me. As if I had done something to her and she had never forgiven me. It has to do with power. She wants to dominate me. I can feel her hatred.
—I can see myself being burnt at the stake, and she made it happen. It's the same energy as when I feel the fear. She is looking at me with the same sadistic laughter, saying "I'm stronger than you."
[Spinning backwards in time with the ISIS technique] —It's in a forest. She is the leader of a group of people, and I came to have a certain influence on some of them. Not that I wanted it or that I did anything to get it. But they were attracted by what I was doing. And she felt threatened. It feels like a long, long time ago. These people are wearing robes with leather belts, and leather sandals. They live in a very thick forest.
—She came to threaten me, but I didn't take any notice because I had faith in the work I was doing. It felt very clear, there was no reason to stop. But she sent some people to get me. They're not really violent or angry. They've got no choice, they're just doing what they've been told. I'm not resisting. I let them take me but I've got a funny feeling. I feel something strange, but with a great inner peace.
—A dirt road, very hot, very humid. They take me to a camp. And then... no need to fight, it's lost. A line of people. I walk in front of them. I start to be afraid. The more I walk, the more I feel something is going to happen. I can hear her laughter. It's the same laughter as the witch's. Like a vengeance. Now she'll have all the power. They take me to a stake, and make me walk around it.
[Crying] —I feel sad, sad. I don't understand why I should die so young, without finishing the work I had started. I can't understand, I feel hopeless. And I feel the hatred of that woman. She laughs while I'm walking up the steps. It's the same laughter. I feel pity, no hatred; more like indifference, and that irritates her even more. They tie me up to a pillar on the stake and she comes near to look at me. It's the same face. She looks like a witch.
—I can see myself burning from above. It's as if a being was trying to lift me up, to take me away from there. There is no pain, but there is something

keeping me there, like an attachment. She enjoys watching me burn. The strength of her hatred frightens me. It's incredible that someone can feel so much hatred – that stupefies me. It's like a magnet. It's not my body, it's her face that keeps me there, her sadistic laughter. Two beings have come to pull me away from there. They are very soft, very friendly. But I'm so sad.
What's the connection between that woman and the presence in your left side? –It's the same face, the same presence. And the same laughter. It's like a vengeance. As if she was not satisfied and was asking for more, as if it had not been sufficient to burn me. She wants to hurt me more.
–The first time I felt the fear was soon after my son was born. I had just cut my husband's beard with scissors. And I started thinking "What if I hurt my son with the scissors?" I couldn't sleep that night. I was afraid I was going to get up and hurt my child. It was her, it was the same presence.
–When she comes into my mind, everything happens extremely quickly, as if I had no control over my thoughts. It makes me afraid of going mad, as if I was not myself any more. It comes when I'm alone. She waits for me to be vulnerable. It's like I'm invaded by her and it gives her great pleasure. It's her revenge. And it's always the same thought: to wake up in the middle of the night and go and kill the children with a knife. Not to strangle them... always with a knife, and blood. When I feel her presence I can always hear the laughter.
Could there be a part of yourself that is attached to her presence? –It's strange, it's as if I was feeling pity for her, as if I wanted to protect her or save her. Maybe it's my protective side.

The client spent ten days watching the witch's presence inside her, according to the method that will be explained in Chapter 14. Then she came back and I performed the regular clearing, after which both the presence and the fear disappeared immediately.

Such examples demonstrate why, when working on oneself, finding an entity is an excellent sign (provided a qualified clearer is available). Suppose the client's phobic fear of killing her children was the result of an early childhood trauma. It might have taken months, if not years, to work through it. Whereas, after the regular ten days of observation, it took twenty minutes to clear the entity, thereby getting rid of the fear and solving the problem.

In this example again we see a client who has been a victim, and a fragment that originated from the offender. If it was the opposite, if the client had been the offender, one could be tempted to see a kind of karmic repayment in being parasitised by the fragment. These two case studies, however, clearly show that a different mechanism takes place: a fragment of the offender is attracted to the victim by a kind of resonance or magnetism, due to the intense emotional connection between the two people. Of course, one could speculate on former karmic relationships between them, and imagine reasons for things to have happened the way they did. Nevertheless, independently of any karmic influence, it remains that fragments are quite stereotyped and programmed in one narrow direction. They mechanically try to satisfy the inclinations imprinted in their astral substance. If they do not dissolve, they wander in the astral space until they find an opportunity to be satisfied.

10.3 Revenge

Case study 10.3 Thirty-two year old woman, herbalist. Exploring an entity located in the left iliac area.

What does it want? –It wants to frighten me and to keep me timid and fearful. It's big. It's anchored in my left side, but it's bigger than me. It stops me from doing anything purposeful. It makes me feel like I'll never do anything with my life.
What does it look like? –Like a big fat person, grossly obese, enormous. Walls of fat. It has a piggy face with gross big lips. And a huge mouth, really revolting, with very red lips, as if they had lipstick. It's got a horrible laugh. Like a little Buddha sitting on a stool. Pink skin. There is something like a cape over its hair. The neck is like a bull neck, because it is so fat. First I thought it was male, but now I see it is female.
–It seems to interfere with my bowels. Lately I've been more constipated than ever before, but I've had troubles in my intestines since I was ten or eleven.
–It [the entity] says I know where it comes from... but I don't know consciously. It's not very cooperative. It feels as if it's paying me back for something. As if I had done something to it.
[Implementing the spinning of the ISIS technique] –I get a vision of a man with the same eyes as the obese woman [i.e. the Buddha-entity]. I see her... and then I see this man, a very distinguished gentleman with one of those big bushy moustaches. He's someone close to me, like my husband, and he has the same sort of neck. But he has a wild look in his eyes, like I've done something dreadful and devastating to him. Oh! Oh! I see the same look in the woman [the entity]. I've really done something disastrous to him [the

husband], like I destroyed him and I've taken satisfaction in it. Oh! Oh! It's like the look in his eyes says: I'm gonna get you. I don't know what I've done to this person, but it seems to be pretty disastrous.

[Spinning] —It's such a dreadful anger. She [the client herself in a past life] wants to kill him, kill the man. She had a baby that was taken away from her. She had a lover and the lover was killed. And then she had a baby but the husband had arranged for the baby to be taken away. A woman came and took it, and she never saw it again.

—After that I've done something to a child. It's revenge, it's been done cold-heartedly. This one is his child, not mine. The child is about six or seven. I see the man with such a wild look in his eyes. I don't know what I've done to him, but I'm amazed at how cold-heartedly it's been done.

[Spinning] —Now I can see the child. He has lots of freckles on his face. It's the man's child. He's being pushed into the water, into a lake.

Who is pushing the child? —The lady. Me. She is very slender. She looks like a refined person. But she is terribly angry. All she wants is revenge. This is the worst thing she could ever do to her husband.

And then what happens? —She drowns the boy. And then I can just see the man again, with the wild look in his eyes. She's satisfied, because she knows it's the worst thing she could ever have done to him. Worse than killing him, because now he has to live with the pain.

In this, as in the former examples, one does not need to talk about karma in order to understand how the fragment got attached to the client. It was attracted to her according to the same kind of polarity as in the two former case studies. These remarks are not made to deny the existence of karma, but to show that in these instances, some more simple mechanisms are at play.

In this case study we notice again how the fragment appears like a grotesque caricature of its origins. A parallel can be established with the way ancient artists, the Greeks for instance, used to depict human passions in the form of masks. This understanding of fragments could certainly lead us to look at our own emotions and desires in quite a different way.

10.4 The flying spirit

Case study 10.4 Twenty-four year old man, masseur. Exploring an entity in the solar plexus area.

What does 'the thing' look like? –It's black, like a black hole in space, it sucks my energy.
Are there any emotions related to it? –Hatred. And fear.
[Spinning mode of the ISIS technique] –I've just landed in a big dark room [i.e. the womb, even though the client won't recognise it immediately]. Very dark. I'm moving in it. I'm alone. The room is huge. Actually I don't even know if it's a room. It's just black. I'm moving in circles, feeling lost. Everything is dark and empty. I feel scared of something. Something is watching me from a distance. There is something trapped inside the room with me. It wants to get out.
Where would it go if it could leave the room? –It would just fly away, straight up, very fast... towards space. But it can't leave the dark room. It's trapped inside it.
What does that thing look like? –A foetus inside a placenta. Very intelligent, though. It's in the room [i.e. the womb]. The room is just like a sludge, a muddy marsh.
Can you feel any connection between this foetus/thing and you? –He knows I'm there; that's why he came. It's a parasite. He's looking for a life-support system.
And where does it find that? –In me. It's looking for vital energy. It looks like a balloon, and inside there is a very small foetus. It's black. And it wants to kill me. It wants to kill me by sucking all my energy. It holds a grudge against me.
[Spinning] –I'm on a big flat piece of land. There are tents, hundreds of tents. And black clouds in the sky. I'm having a fight with another man. I have a huge sword, and I cut him in two halves. He looked like quite a big man, but I'm even bigger than him. Much bigger than I am now. He was my friend. We had an argument. Some kind of power fight. He was the ruler of the camp. And now I am.
–It was a master blow, I've cut him in two halves, across here [showing the navel area]. And then I saw a spirit flying out of him into the sky. I'm completely amazed.
–Now the people are coming out of their tents, and they're scared of me. They look like Mongols or people from Tibet, that sort of facial appear-

> ance. And I just can't move, so surprised by what I saw, staring intensely at the spirit flying into the sky!
> *Where did the spirit go from there?* —It waited. It was waiting for me, hiding somewhere in the mud. It was making a foetus in the mud. And then it came into the dark room.
> [Spinning] —I can see myself hugging this man. He was my best friend. There was a lot of warmth between us before the fight.
> *Could there still be some of this warm feeling between the foetus and you?* —Yes. Actually I'm surprised at this warm feeling. It's not only me who likes it, it also likes me.

Note that the client's last words reveal how it is not only the grudge, but also the deep friendship between the two men that created the link, the fragment's attraction to him.

The way the fragment found its way into the womb is wonderfully described. As in many other cases, it made me marvel at how coherent clients' observations can be. Virtually none of the clients presented in this book had ever heard about entities before going through ISIS. No theoretical explanation had been given to them before starting the practice. If the topic had been mentioned in conversation, they probably would have very much doubted that entities exist. Yet during the sessions, they took what they saw quite naturally. They related precise and meaningful details, exactly as if they had read this book!

10.5 The Roman deserter

> **Case study 10.5** Thirty-nine year old woman, chef. Exploring a shadow discovered in her left hip in ISIS.
>
> *What are you seeing?* —Picture of being on a chariot. I shouldn't be there. I should be with the people. I should be fighting with them. But I ran away and I was wounded in the hip with a spear. They've just left me there to die. It's a soldier, left dying in a chariot. He [i.e. the client in a past life] looks like a Roman soldier. Running away from the other soldiers. A mercenary, but the people I'm supposed to fight are people I've known since my childhood. So I run away and they throw the spear at me. [In this Aus-

tralian life] I've always been a bit of a coward. I've always been frightened, terrified by any form of violence.[1]
—I don't have any right to have the good things because I should be dying with them, instead of running away. I grew up in this village and I became a soldier in order to be able to see the world. I never expected that I would have to come back and kill these people. I feel so ashamed of what I have done. I should be with the people of the village, dying with them. And in this life I've always identified with people of poor countries. I've always felt that I belonged with them. I just feel so ashamed of what I've done.
What's the connection between the shadow and the soldier? —The shadow is there to punish me for what the soldier did. I should never have left these people. So the shadow is hiding behind a screen of weakness and making mistakes. I can never be sure of what's right and what's wrong. My thinking is a bit confused. I take other people's opinions, because it's easier than thinking my own. The shadow wants me to be overpowered and treated badly. It wants to put me in prison, to limit me, to limit my expression. My parents used to lock me inside the house. I was raped several times, and the shadow attracted it.
Where did that shadow come from? —It was in the soldier. The shadow was the soldier. The soldier wants me to pay for the mistakes I made when I was the soldier.

This case study presents a new mechanism: a client parasitised by a fragment that broke off from her own astral body in a past life. With guilt as a connecting link, she is now tormented by an entity coming from herself in the past.

In this particular example, the borderline between samskara and fragment-entity is far from being clearly delineated. Samskaras are imprints coming from traumatic and intense experiences of the past, collected either in this life or in former ones. A track is left in the astral body due to the experience, and carried from one life to another. A fragment, on the other hand, is a piece that breaks off from the astral body after death, and becomes attached to the astral body again in this life.

In theory, the distinction between samskara and entity may sound subtle. In practice, however, there is one major difference: an entity can be

---

[1] In ISIS regression, it is quite common for clients to alternate between the first and the third person while talking about themselves in a past-life episode.

cleared in twenty minutes, a samskara can't. During clearings, clients can see the entity suddenly moving up, out of their body. If one could clear samskaras like this, anybody could be enlightened within a few weeks.

Also notice how the client perceived that there was a direct connection between the shadow and her being raped several times. It is not uncommon for clients to discover that their entity has attracted various negative circumstances to them, including accidents and other forms of physical violence. More than once I have heard clients report that their entity had made them black out for one critical second while driving, thereby nearly causing a car accident. In other cases, entities appear to modify the client's behaviour in a way that provokes violence from other people. In some more puzzling cases, clients perceive that their entity directly manipulated other people to attack them.

10.6 The tight collar

Case study 10.6 Twenty-four year old man, working with computers. He was strongly committed to an ascetic form of yoga, and got up at four o'clock every morning to meditate. He was strictly vegetarian, refused to have a girlfriend, and saw any commitment to the material world as unspiritual and dirty. Even though he was under the impression that he was doing all the right things, he felt completely inhibited and smothered by self-destructive emotions.

What does it look like? –It looks like a red cloud, almost like a collar around my neck, pressing, strangling me. It's almost as if there is no space to breathe. It's pressing harder and harder, it's a pretty malignant thing.
Could there be something like a presence attached to it? –Strangely, yes. It almost seems to have a life of its own.
Are there some foods that it enjoys? –Cheese, and anything sour.
What happens to the red cloud when you have these foods? –I think it gets stronger. And a little bit bigger. And the presence on my neck gets a little bit more intense.
What does it want? –It wants spiritual things, like meditation.
What happens to it when you meditate? –It gets more intense, it strangles me more. And it has more control over me.
What else does it want? –It wants success, spiritual success. It wants to be in control. It controls me, it puts pressure on me.
[Spinning] –Two year old, in the street with my mother. The cloud was already there, but smaller, attached to the front of my throat.

> [Spinning] –A dark space...
> [Spinning] –I'm sitting down close to a window and writing. It's an elderly man, writing a letter. He's got wrinkles on his hands. He's like a priest, or something which has to do with the church. He thinks a lot. He's not happy. He feels tense. He's got lots of expectations about himself, ideas about what he should be. He thinks he should be more than he is. He is wearing a priest's collar.
> *What's the connection with your red cloud?* –My cloud has to do with his expectations, the high idea of what he should be. When I think of this collar it feels really familiar. It does not do good but it feels right. The collar represents a lot to him. The collar gives me feelings about how I should behave. It's like an expectation. The cloud is a restriction. His collar was too. It's like I can feel the presence of the priest in the red cloud. When I see the cloud, and when I see the priest, the feeling is really similar. It's the fanaticism of the priest that feeds the cloud.
> *What's the connection between you and the priest?* –The priest is me before. And now the priest is in the red cloud.

At last we meet an entity that likes meditation—unfortunately, not for the right reasons! Again this example shows a fragment coming from a past life. The situation is complex: the entity is intricately mingled with the client's samskaras (i.e. psychological dispositions). Dealing with the samskara, that is, undergoing some kind of psychotherapy, would certainly not be enough to get rid of the fragment, for entities are particularly tenacious: I have observed in hundreds of cases that they almost never go, unless the proper clearing process is implemented. However, it is not enough to clear the entity to solve all the client's problems.

Entities are like a lock on a door. By clearing the entity, one gets rid of the lock. This is not enough to open the door, but still it makes a big difference. In many cases where clients have had a combination of an entity with a major psychological blockage, I have observed that as long as the entity remained, the clients were unable to make a move towards resolution. Once the entity was cleared the problem did not disappear, but the clients often started resolving and releasing the corresponding samskaras.

This mechanism should be remembered when a client is completely stuck in a regression process or some other form of psychological work. It could be that an entity cements the blockage, in which case having the entity cleared can be a quick and powerful way of facilitating a shift. Yet using entities as scapegoats each time someone's development is hindered would be ridiculous, for there are hundreds of ways of blocking and resisting psychological shifts which have nothing to do with entities.

10.7 The Indonesian woman

Case study 10.7 Thirty-eight year old man, politician. Exploring a dark cloud on his shoulders, like a weight that drains him and blocks him psychologically.

What are you feeling? –A real heaviness throughout my body. A physical heaviness. Like a menacing dark presence. There is this image of the elderly Indonesian woman in her decline. [Prior to this session, the client went through several regressions that led him to identify this Indonesian woman as a past life of his.]
–She looks tired. And when I see her, when I feel her presence, it makes me become aware of my back. She's got dark skin. She wears a skirt made of a kind of grass.
Are there any emotions or feelings associated with her presence? –She is not really feeling anything. She is oppressed. It's like a weight, a heaviness in her back, in her shoulders. I don't think she knows what she is oppressed by. It's like a presence on her back. It's like a clinging for life. It wants her life essence. I see her as a vibrant young woman and it wants that. It clings to her back because it wants her lovely energy.
[Spinning] –Her husband was jealous and killed her lover in front of her with a machete. And she was mortified by that. I see the lover as he was dying. There is a sense of him not wanting to go. She was terrified.
–The weight on her back started after that. The weight is the lover's presence. It's like him but it's not exactly the same as when he was alive. When he was alive, his energy was OK. But the good aspect of his energy when he was alive does not seem to be with him any more. It's become dark. It just wants to drain her. It's not loving any more, it's abusing her.
–The Indonesian woman declined to the point that she was so overpowered by it she became an outcast. Her husband did not want her anymore and she was too exhausted to go out. She died alone.
–I can feel it in my own body, the same weight, the same presence.
What does it want in your body? –My good bits, my positive energy. It enjoys the life force vicariously. It saps it.
Are there some foods that it enjoys? –Heavy fried foods. Maybe sweets too. Heavy foods.

> *Does it interfere with your sexual life?* —It's more after sex than after love. It likes sex because sex is life. It gets nourished through it. It wants more, always more.

This case study reveals a new mechanism: a client carrying the same entity from one life to another. In terms of our model of subtle bodies, one can easily understand how this may happen. When fragments wander in the astral space, they are attracted to what feels familiar to them. If, in a past life, an entity had been attracted to you for a specific reason (not only to drink beer or eat chocolate) and was not dissolved after your death, it is logical that it may keep on looking for you and wait for your return.

10.8 The need for caution with past-life experiences

Before closing this chapter on past lives and entities, one reservation should be clearly expressed. The fact of discovering a past-life episode while exploring an entity should not be regarded in itself as an assurance that the client actually lived that particular past life.

Fragments are not very clever. They have quite a low level of organisation. While wandering in the astral space, all they want is to repeat the emotions imprinted in their substance. Like attracts like. Fragments look for people who can satisfy their cravings.

Suppose a fragment carries a desire for vengeance against a particular person. It will wander in the space, looking for that person. It remembers the quality of vibration of the person, and therefore looks for the same quality of vibration. However, that does not mean it will end up finding the same person. It just means it will end up in a person who has the same quality of vibration. Remember, fragments are basically stupid. All they can do is repeat the same emotional reactions endlessly. We can't take it for granted that they will be specific enough to find the same soul again. They may well be misled and attach themselves to someone with similar astral features and psychological dispositions.

Clients may be confused because of all the memories they discover about that past life. They may well end up believing the story to be a past life of their own. But in reality that episode has nothing to do with their own past, it is just a life story imprinted in the fragment. It is someone else's past life that the client is discovering. In certain cases, a lot of discernment is required to be able to know what comes from what.

CHAPTER 11

CORDS

11.1 A typical cord case

Case study 11.1 Twenty-two year old woman, salesperson. The first thing she told me was: "I don't believe in entities but I can clearly feel an energy that wants to leave me and that does not belong to me. It feels separate. It's related to my mother's energy. [The mother was still alive.] It's been living in me but now it's time for it to go." Then she told me that this foreign energy expressed itself in an extremely violent way, by creating anxiety attacks. It sent to her mind visions of "horrible things" happening to her, such as stabbing herself in the guts.

What are you feeling? –It's like being stabbed in the belly [below the navel]. It's a very small spot, like a point.
Is there the feeling of a presence attached to it? –No, it doesn't feel like a presence. It's not willing to come to the surface at all. It keeps connecting with my head and telling me it's not there. It's trying to trick me all the time. [Flow of tears] It's like despair, lots of pain and despair. It's like the hopelessness of a totally sad reality. And it's frustrated and angry at being so unhappy. It just wants not to exist. So I don't concentrate on being here, I'm not really in the world. It might cause me to be destroyed, to commit suicide.
Where would it go if you died? –It would stick to me, to my consciousness. That's the real joke: it would be just as unhappy, but not in a physical form. [Spinning mode of the ISIS technique] –That's when I was three years old. The energy is already there. It gives me the feeling of not being safe, of not feeling grounded in my body, of being physically insecure. Fear, fear fear... seeking attention all the time, needing to be reassured by another person. [Spinning backwards in time with the ISIS technique.]

Cords

> −It's dark. It's in the womb. The energy is already there. It feels connected to my mother. It grew physically here because she was so unhappy. I don't like being in the womb. I'm surrounded by this unhappiness, simply because she is so unhappy herself. Actually this thing is like concentrated unhappiness. And as I'm growing, it's becoming me as well.
> *Does it have a shape?* −It looked like an umbilical cord, and it was growing in my stomach. The cord is still there, even now; [flow of tears] but now it's just black and horrible. It's attached to my navel on one end, and to my mother on the other end. That's why I can feel her energy all the time. I can feel her presence any time, on the other end of the cord. It's like I can hear her all the time complaining about how unhappy she is.
> *What's the cord made of?* −It looks like dead flesh and dried blood, it's just horrible. I'm very aware that my mother knows it's there, and it's terrible for her to see it.

As soon as the clearing was performed, the client stopped feeling surrounded by her mother's presence, and her anxiety level dropped dramatically. Coincidentally, the young woman's medical history before the discovery of the cord included endometriosis, pelvic inflammatory disease, and fifty cysts on both ovaries. The cysts had been operated on but, due to a bad scar, a second operation was needed one week later (which should immediately lead one to suspect the presence of an entity).[1] During the ISIS sessions, she repeatedly commented that the black cord had a similar feeling to the pain of her various abdominal illnesses. In other words, the cord probably played a key role in the genesis of all these physical troubles.

11.2 What is a cord?

In terms of subtle bodies, how can we understand this cord? The umbilical cord that links a mother with her baby not only carries physical substances but also life force, meaning etheric energy. The umbilical cord is therefore not only a physical structure, it also has an etheric layer attached to it. In Chapters 6 and 7, we described how getting rid of the physical foetus after a termination, or of the physical placenta after birth, was not always sufficient to eliminate their etheric counterpart. Similarly, the

---

[1] As discussed in Chapter 8, 'How does one catch an entity?', any surgery, especially when performed under general anaesthetic, is a high risk situation for catching entities. Complications after surgery do not necessarily indicate the presence of an entity, nevertheless they should lead one to suspect it.

cutting of the physical umbilical cord after birth does not necessarily mean its etheric counterpart is severed.

During the first years of a child's life, intense exchanges are maintained with the mother on various levels, from etheric energy to affection and love. The state of dependency of the child on the mother is perfectly normal, and there is not necessarily anything wrong with the persistence of a cord for a few years. However, if cords remain for too long, they prevent the children from gaining their psychological autonomy. They not only create a twisted relationship, they block the children from undergoing their natural development and reaching maturity. Later on the mother may be unable to let go of the child (who is no longer a child), the child may remain abnormally linked to the parents, or another distorted way of relating may develop.

From my observations, a number of pathological relationships between parents and children are actually due to such cords. However, until now, very few people have had sufficient perception to realise their presence and their nature. Through conventional means of exploration, people only become aware of the psychological side of the relationship problem. If they try to deal with it, it is only through psychotherapy, exploring emotions and learning to deal with them better. But as long as the etheric cord remains with its inappropriate transfers of etheric energies and emotions, it is quite difficult to reach a real resolution of the problem.

From my experience, many cords can be just as stubborn as entities. As long as they are not cleared in the same way as entities, nothing can remove them. Thus people can work at exploring their problems with their parents (or children) for ten years or more without ever coming to terms with them.

11.3 Can a cord be called an entity?

Cords certainly share a number of characteristics with entities. They drain the client's life force. Through a cord, the client receives a number of parasitic emotions from the parent, or the child. A cord can also be responsible for cravings and a few other mechanisms we have described for entities. Moreover, a number of cords are very difficult to cut and get rid of without a proper clearing, similar to those implemented on entities. However, if such a clearing is performed, as with entities, the cord disappears immediately and does not come back.

There is another good reason why cords should be dealt with in the same way as entities. Suppose someone was clever and strong enough to be able to cut a cord without a proper clearing. The result would be a fallout

Cords

of etheric and astral bits and pieces, which would be as toxic to the cord-cutter as to the client. Furthermore, a partial clearing that left bits of the cord would make the situation even more difficult to resolve in the future, as is the case with entities.

Cords, however, unlike entities, do not have a consciousness of their own. Therefore they do not create the feeling of a presence. For instance, a cord by itself cannot generate voices in your head (even though it can cause you to be invaded by the thoughts of the person on the other end). A cord is not an individualised consciousness, but a link with someone else's.

In the case study we have just seen, the cord was not a foreign energy that came to be attached to the client one day, as entities are, but the abnormal persistence of a normal structure. This last remark does not apply to all cords, as we shall see later. Yet all cords are more or less self-generated by the two people they link, even though one of the two has often played a much greater role in its establishment.

In conclusion, cords are not entities, but the best way to deal with them is to treat them like entities. This implies exploring and watching them (see Chapter 14), and then clearing them like entities.

As far as healing is concerned, the results of clearing a cord are the same as those for entities. It is not enough to clear the cord to heal the relationship and the emotions attached to it. A cord is like a lock on a door. It immobilises the problem, it makes it unsolvable. As long as the cord remains, it is extremely difficult for the client to deal with the corresponding emotions. Once cleared, a lot of work still remains to be done. However, everything becomes much lighter and the client can start moving towards a resolution of the inner and outer conflicts.

We are now at a time when a significant proportion of the population is awakening to the perception of the etheric layer. Consequently, many will become aware of these cords. Note that cords are not new! The new element is that many people are at the threshold of perceiving them. I foresee that the work on cords could gain great popularity in the coming decades, for it is a remarkably efficient way to improve parent-child relationships, which, needless to say, account for a good part of modern neurosis.

Case study 11.2 Twenty-six year old woman. Exploring a love-hate relationship with her mother.

What are you feeling? –There is a cord, like an umbilical cord. It looks like twisted skin, like an umbilical cord looks. It's pink.

What's passing through the cord? —Blood... and anger and hatred and bitterness... and envy... and grief. It's been there ever since I was born. As long as it is there, I will never be free. I'll never be able to have a relationship with a man. It takes all my energy. I have no power left.

What's on the other end of the cord? —My mother. She is sucking my energy. And with me on the other end of the cord, she is never alone. She's always been scared of being alone. She thinks that she'd die if she were alone.

Is there a part of yourself that enjoys having the cord? —Yes, because I'm not alone either. There is a part in me that believes that I might die if I were alone.

Case study 11.3 Forty-one year old man, physician. Here is his report of a forty-two-year cord story.

—It was insidious. I just couldn't believe that this thing had been here all these years without me seeing it. But it felt very real; subtle but tangible. It started from below my navel. And on the other end there was my mother. I couldn't see exactly where it was attached to her body, but her presence felt very real. It's like someone was on the phone with you and you forgot to hang up.

—There was the feeling of a constant energy drain through this cord. When I was a child, my mother was very possessive. And through the cord, she could keep on holding me even after I became an adult. She may not have been aware of it, but she did not want me to grow up, she always wanted me to remain her little boy. So the cord was like a plug, limiting me and destroying my self-confidence; and draining my energy.

—Even though I couldn't see it then, it was because of the cord that I felt so much anger toward her. I could sense she was doing something like that to me, just I didn't know exactly what. And that was making me frustrated and angry. I hated her. When I was a teenager, I had huge fights with her all the time, I even coldly considered killing her a number of times. I couldn't get rid of the plug, so I wanted to get rid of her by any means I could.

—Unfortunately, once I left home, the plug remained. I had wanted to get away from her for so long, and now I discovered that I couldn't forget about her. Not that I missed her, but she was still present in my mind too often, and I couldn't find any reason for that. I thought it was just resentment, but it was the cord.

> —As soon as the cord was cleared, a part of me felt so much lighter... I couldn't believe it. Also I started feeling less angry at my mother. And I could start building up my self-confidence, and looking for my power. Now I still don't have much esteem for my mother, but I don't hate her and I don't fight with her any more.

11.4 Cords created by relationships

So far we have only considered one type of cord: the abnormal persistence of the umbilical cord. In the ISIS state clients also describe other cords of a similar nature, but originating from relationships other than that of parent-child.

> **Case study 11.4** Thirty-one year old woman, film producer. One year before consulting me, she ended an eight-year relationship. Even though she was the one who terminated the relationship, she was unable to let go of her former partner. She kept on oscillating, one part of her knowing that the relationship was wrong, the other wanting to hold on to it and start again with the same man. When she came to see me she was heavily distressed and depressed. Here is what she discovered while in the ISIS state.
>
> *What are you feeling?* —I feel something... like an umbilical cord.
> *Made of what?* —Blood vessels. It's one big cord, like a big blood vessel. And it's linked to him.
> *To which part of his body?* —Same. The navel. I feel enclosed within him, within his space. All the time.
> *Is there anything transiting through the cord?* —It's like energy, love, blood, and air, vital force. It's like he [the former partner] has made the decision to leave me with his mind, but the cord is not cut.
> *How did this cord come to be there?* —It developed with time when we were living together, from the love we shared.

When taken into the ISIS state of expanded perception, a number of clients suddenly realise that they are linked to someone by a cord. The person on the other end is nearly always someone with whom they have had an intense emotional association.

Clients describe this cord in various ways, the most common being an 'umbilical cord', but not necessarily linked to their belly. The second most frequent place where the cord can be attached seems to be the area of the heart, either on the left side of chest, or in the heart centre in the middle of the chest. Sometimes they describe the cord in nice terms, such as "a beautiful stream of energy linking our hearts" or "a star-lit rainbow linking my umbilicus with my daughter". In other cases its appearance is much less appealing: "It looks like dead flesh, it's sick" or "It looks like an old rotten sausage, all black, revolting and smelly."

As with entities, it must be emphasised that I never mention the term 'cord' to my clients unless they see one themselves first. Nor do I suggest the possibility to them by some kind of indirect question. In all the examples I have presented here, the clients were completely unbiased, since the concept of an umbilical cord binding you to your boyfriend is quite foreign, to say the least, to the usual western mind. Yet how much energy is spent on relationships in this western world! Cords therefore appear as a crucial topic, for without them one can't really understand what is taking place in certain relationships in terms of energy.

Before exploring their mechanisms in more detail, let us make one point about cords very clear: they are not a fancy of the mind. They are neither an illusion nor a kind of hallucination, nor even a subconscious archetype dreamt by the client. They are very real and tangible, made of etheric and astral substances. The ISIS techniques have nothing to do with hypnosis or suggestion. Moreover, once discovered, it is not only during ISIS sessions that clients can see or feel their cord, but any time during their daily activities. At the time of the clearing, they can often see the cord being dislocated and moved up into the light. As soon as the clearing is completed, they feel significantly better. As with all entity-related structures, psychological work still has to be completed afterwards, but it is significantly eased by the clearing.

From the point of view of subtle bodies, a cord is a kind of etheric-astral pipe, through which energies and emotions circulate more or less constantly. On the astral level, emotions are not abstract things but shapes and waves made of various astral substances. Any emotional exchange between two people creates a flow of astral energies. However, a cord is more than just a flow, it is a permanent circulation linking the two. It does not only transmit emotional waves but also life force, i.e. etheric energies.

When a man and a woman live together, their constant emotional interaction is an initial factor favouring the creation of a cord. Sleeping in the same bed for years adds to that by creating an intermingling of etheric energies. Sexual intercourse is also a deep exchange, in which a lot of life force is involved. If a woman has a child with a man, it means that for nine months she carries his chromosomes, and therefore the personal blueprint of his etheric life force—and that too creates a strong bond of energy. Just by becoming pregnant, even if a termination follows, a woman carries a man's chromosomes as long as the embryo is inside her, which leaves an imprint in her energy.

It would be fascinating to study the marriage rituals of various religions or civilisations in depth, to determine what kind of spiritual or even occult forces are involved in them. What is obvious is that they are all designed to reinforce the bond between a man and a woman, and they are trying to do so by imprinting some kind of spiritual influence on them.

None of the binding factors I have just described is new. What is new is that a large percentage of marriages end in divorce. A growing number of people who have spent years establishing extremely deep links between each other choose to separate. Not knowing how to untie the energy links between them, they go through difficult and painful times. My position is of course not to say that there is anything wrong with divorce, but that a lot of suffering could be avoided if these energy ties could be recognised and cleared.

Case study 11.5 Thirty-eight year old woman, mother of two teenage children and in the process of divorcing her husband, who left her seven months before this session.

What are you feeling? –The cord hurts. I know it's stupid because I don't want to be with him [the former husband] any more. I don't even feel any love. I'm not angry at him anymore. But each time I see him, it hurts in my belly, around where the cord is attached. It's like a blow, it hits me, and I lose my confidence. And then I become emotional. It's stupid but each time I see him it takes me three days to get over it.
What does the cord look like? –A kind of jelly. A mixture of blue and pink. Sometimes there is like an electric current going through it. I know he has the same, only he is better than me at pretending it's not there.

11.5 Other cords

Let us now have a look at a few different types of cords.

Entities

Case study 11.6 Forty-nine year old woman, medical secretary. The following exploration was conducted more than five years after the death of her mother.

–I could see a cord floating in the space. I followed the cord. I spun back a long way. I crossed a barrier into another space. It was denser and I was aware of vague presences there. I could not clearly define them though, it was as though they were muffled. There was pain there, and loss and sadness –all the qualities I feel in my space.

–The cord led me to a presence. I felt it was my mother and I was afraid to see. But she wasn't really there. It was just a shell of her [=a fragment]. She'd gone off somewhere else. I didn't know what this meant. There were lots of other cords in that space, quite a network, but they belonged to other people. They weren't mine.

–Through the cord, my mother is feeding off me in much the same way she did when she was alive. I can feel this desperate need and it isn't mine, it's hers. It's like my mother is grasping and pulling at me because she doesn't want to lose contact, and of course I love her and I don't want to hurt her.

–The cord is attached to the upper outer quadrant of my left breast. It reminds me of the way a cord is attached to the placenta. It looks exactly the same. I can physically feel a real drawing feeling from the cord. It is drawing energy into itself.

–Is my depression hers? No, not altogether. I have had my own depression but this close connection has enabled me to feel her pain and sadness too, and the whole thing has got mixed up together. It's a bit like my mother's pain has been my pain and my pain has fed into hers. Pain was one of the things we had in common –pain and depression.

[After the clearing of the cord:]

–When the cord was being removed, I could feel just how deeply it had penetrated. It was a big, fat, healthy cord that had had a lot of use. Its roots were really very deep, particularly in one spot in the heart. It covered a very large area spreading right into my axilla. When it was removed there was a big hole/space left where it had been.

–The day following the removal I felt a real sense of freedom. It was as though I had been captive all my life without knowing it and then suddenly I was free.

If we analyse this case study in terms of subtle bodies, we see someone who used to be linked to her mother by a cord. This cord could either

Cords

have been the continuation of an umbilical cord that gradually moved from the navel to the heart over the years, or a new cord that started in her heart because of the way the two women related to one another.

When the mother died, the cord remained. The client was no longer attached to her mother, but to a fragment. After death, her mother's astral body underwent the usual shattering. However, the astral part that used to be connected with the client through the cord remained hanging around in the space, as if nothing had changed. The fragment did not have to wander in the space and look for someone to give it satisfaction. It could just keep on getting it through the long established cord. All the negative exchanges that used to parasitise the client remained virtually unchanged. This shows that the death of the person on the other end of a cord is not necessarily enough to break the connection.

It also introduces an important concept: a cord does not necessarily link you to a person, it can also be attached to a fragment.

Let us now look at a case in which another type of cord is involved.

***

Case study 11.7 Twenty-seven year old man, musician. Exploring a cord implanted on the left side of his navel.

What are you feeling? –It feels like a whirlpool. There is a spiral of energy whirling up the pool. It's a murky grey black sort of colour.
What's on the other end of the cord? –A whole space, a dark space, with all these little beings, not unlike gremlins. They live in an orgy of self-indulgence, whatever form of pleasure they can get through me. They thrive on all the base emotions. Now I can see them. They are always waiting for the next feed, waiting to be satisfied by my next burst of emotions. And as soon as it happens, it's like feeding time at the zoo.
–Around this cord it's like a video library of all my sexual experiences, all the women I've made love to. It gives the little beings a thrill. The more junk food I eat or the more alcohol I drink, the more active they become as well. Then their energy gets out of control. Grotesque little beings, involved in a permanent orgy, they never stop. Each time I indulge in any form of pleasure, they get completely frantic and it starts generating an orgy. The other night I had a dream full of sexual images and when I woke up I could see it was coming straight from them.
Are there sometimes voices related to the cord? –Sometimes the little beings send me a "Have another drink", or "Eat more of that stuff", or "Talk to that girl over there". They have a lust for all these things. They can only get their satisfaction through me. They get it from the physical plane, and

> they need a connection with it. They're in a space where all these things can't be found. They can't satisfy themselves from there, but they know where to find what they want. It's just a question of making a connection through someone. And they do that through me. They make me feel like I am an addict, hooked to the euphoric alcoholic feeling, or what I get from women. So I hold on to the little beings so I can satisfy my addiction.
> *Why do you need them?* –Because they heighten the feeling and I enjoy that.

In most cases, even if an addiction is directly related to an entity or a cord, a clearing won't be enough to get rid of the addiction. However, as long as a client remains connected to a lower astral plane like this one through a cord, it makes it infinitely more difficult to resolve the problem.

From what I have observed in dozens of similar cases, it seems quite common for people suffering from an addiction to alcohol or narcotic drugs to have a link with a lower astral plane, whether it is through a cord or not. The same sometimes applies to people who have an extremely intense sexual life. As the client clearly described in the last example, the beings in these spaces are craving sensual enjoyment of various kinds, but are unable to get it where they are. They need a connection with the physical plane. In the astral, like attracts like. These beings are therefore naturally attracted to people whose life is centred on intense sensual enjoyment. The regular consumption of a drug, from alcohol to heroin, also makes a person particularly attractive to them.

Let us finish this chapter with a more sophisticated cord that seems to reappear from one reincarnation to another.

> ***Case study 11.8*** Thirty-six year old man, university lecturer. After separating from his wife a few months ago, he finds it terribly difficult to let go, not of the wife, but of his seven year old daughter. He keeps on seeing her twice a week but his suffering does not seem to subside with time. He realises that he is clinging to his daughter in an unhealthy way whenever they are together. Even though he is quite aware that something abnormal is going on inside him, he doesn't seem to be able to do anything about it.
>
> *What are you feeling?* –Pain, pain in the heart. It's like a string, or a cord attached to my daughter. It starts there, in my heart. It's as if I didn't know if this part of my heart was mine or hers, as if I could not differentiate. And I can feel something really heavy about this part, something old. Something not fresh, not alive. It looks like an old chicken leg, all shrivelled, with

scales. That's the cord that links us. It feels wrong, it doesn't flow. It shouldn't be there, it's unnatural and heavy.

Is it related to the pain in your heart? —Yes, as if the cord got twisted around my heart. It squeezes my heart, it smothers it. I'm sure it's connected with my asthma. As if the cord took over and wanted to control everything [entity feature]. As if my daughter and I were both prisoners of it. Each of us is linked to the cord. It's like there is a part of her inside me. A part of her heart. Right here, in the middle of my chest.

[Spinning] —Two people working with flowers and leaves, like botanists or explorers. She is my assistant. There is a beautiful link in the heart. A strong bond. I think I was in love with her, but I was older than her. At least ten or fifteen years, and I did not want to tell her that I loved her. The cord was already there, but it was not so rigid. It looked like a pipe with scales, but not like a chicken leg, much more flexible. It started in my heart, but I see it going into her head, not into her heart. It was already something heavy, more like a chain than anything else. It prevented me from communicating well with her.

[Spinning] —They travel a lot together, but their way of relating remains the same. They stay together for a long, long time. She has a lot of devotion to him, but it remains unspoken.

[Spinning] —He is the one who dies first. He drowns, in an accident. The cord breaks, and remains with the woman. She leads a solitary life, she still feels the attachment and it prevents her from opening to someone else. They each have a bit of the cord, but the cord became elongated when he died, and broke.

—For many days, after my daughter's birth, I was in a fantastic state. I wanted it to be perfect love. It's as if the cords had been linked again. I thought we were complementing each other, but in reality the cords were being knotted again. But now the cord looks old and shrivelled. It feels wrong, it has to leave me. It's like an old boat anchor, heavy and rusty.

—And there is a bit of me in her heart. That had already happened with the botanists. As if a part of me had passed to her through the chicken leg, and a bit of her had passed into me. That's why I reacted so strongly when she was born. I felt full, with her, I felt very strong, invincible, irresistible through her.

In terms of subtle bodies, in this example we see more than just a cord between father and daughter. The two have exchanged bits of their astral bodies, not unlike the way players exchange shirts after a football game. Consequently they remain linked, which results in a lot of suffering.

One can easily imagine that such an intricate connection did not facilitate the psychological development of the child.

In a number of cord cases it is difficult to discern whether there is only an exchange of energies and emotions between the two people, or if they have actually exchanged parts of their astral body. For instance, if you carefully read case study 11.1 again, you will see that such an astral exchange could well have taken place between the client and her mother.

The mechanisms by which many mothers tend to purge themselves during pregnancy and send some of their astral parts into the foetus is certainly a facilitating factor for such astral exchanges to take place. From what I have observed, I consider that many difficult problems parents have with their children are due to a similar mechanism. Part of the child is still in the mother, and/or part of the mother is in the child, with a strong cord binding the two. However, having a fragment of one's astral body trapped in a child or a parent is not the only way of having an astral bit missing. This is the topic of our next chapter.

CHAPTER 12

MISSING BITS

12.1 What are missing bits?

> **Case study 12.1** forty-six year old woman, unemployed. She suffers from chronic fatigue, headaches, and lack of motivation. She would like to get back into the work force, but is struggling against a huge inertia, as if she simply couldn't be bothered trying to find a job.
>
> –Sad. About eight years old. I'm in a rage. Locked up in the bathroom. The housekeeper locked me in the bathroom because I've been naughty. I'm crying and kicking, enraged. And then suddenly I stop fighting. Something disconnects. I become quiet, but dull. Something has left me.
> –There is a part of me that left me while I was in that bathroom. It went out of me. And then the part was not with me any more. I was a very determined child until then. I lost my strength of character. I became like a blob.
> –It's been so long since this part has been with me. It's just not part of me any more. It's like a part of a jig-saw puzzle. I can see it around me but it's not in me.
> *What does the missing bit look like?* –It's made of colours. There is some purple, and some yellow. And it's linked to my heart by a cord. It feels soft and warm like a pillow. It's something I need. Otherwise I can't really be myself.

In a way, 'missing bits' can be regarded as the opposite of entities. An entity is a foreign energy that comes to be attached and sticks to you like a parasite. A missing bit is a part of yourself, a part of your own energy that was ejected for one reason or another. To use the words of a client: "It is like a missing part in your own jig-saw puzzle".

Entities

From the point of view of subtle bodies, a missing bit is a part of your astral body that came to be separated, exiled in some distant astral space. It remains linked to your astral body by a cord, similar to those we described in the last chapter.

The astral body is made of prominent parts involved in performing the basic functions of life, and also of more subtle and spiritual parts. One doesn't lose prominent bits of one's astral body easily. A significant trauma is usually needed for the disconnection to take place. The trauma creates a breach in your energy, a temporary collapse of your etheric and astral defence mechanisms which allows the bit to drift away—just as entities need a breach in your energy to be able to penetrate your system.

The astral body is the layer of thoughts and emotions. However, in many cases one loses much more than a bit of one's mental capacities when an astral piece departs. If we go back to our fourfold pattern of subtle bodies (physical, etheric, astral and Ego), we see that the etheric is an intermediary between the physical and astral bodies. The etheric transmits astral impulses into the physical body. It acts as a link. Similarly, but on a higher octave, the astral body acts as a bridge between the Ego (i.e. Higher Self) and the etheric and physical bodies. If a key bit of your astral body is missing, the result may be that your Ego cannot be fully present in your etheric and physical bodies, and therefore in your usual mental consciousness. In other words, the Ego can't fully take part in your daily life. The result is an 'absence'— whereas entities add their astral presence to your own.

Common symptoms due to a missing bit are: fatigue, lack of energy, lack of stamina, lack of motivation, lack of purpose, lack of 'inner flame', lack of drive, lack of self-confidence and courage, lack of joy... a lot of 'lack'. The clients have the feeling of living an empty life, of walking through their various occupations like zombies. Apart from these central symptoms, various psychological and physical disorders may develop, according to the dispositions of the clients and their life circumstances. For instance, I have observed certain obstinate skin disorders such as psoriasis, that resist virtually any conventional forms of treatment. Of course only a few skin diseases are due to a missing bit, and many missing bits never create any physical ailment.

12.2 The return

Case study 12.2 Forty year old woman, social worker. The following passage describes the return of a missing bit, possibly lost at the age of five, when she was completely terrorised by a cocker-spaniel barking at her. It had taken a lot of work, exploration and clearing to get in touch with the missing bit.

–There is something creeping on the top of my head, trying to get in, something that's dark but not scary. It's afraid to come in. It's like a dark cloud, like a little spirit. It doesn't quite know whether it should come in or not, but it's drawn to come in. It's very tentative. For some reason my body wants it to come in. There seems to be a struggle, because it's not used to being there. The struggle creates tension in my head. I know that it's a good energy but I don't know how to draw it in, and it's exhausting. It creates pain in my head and my whole body when it tries to come in.
–I feel a bit shaky. I can feel the struggle. It's entering my body through my arms too, not only my head. It's like it must come in and take control. The body needs it to take control, but it's scared of taking control. And the body is scared too of being controlled. It's very energising. I feel so much energy, I want to scream with it. It's nearly too much for my body, it makes me feel shaky. I'm not sure if I'm ready and I feel overwhelmed. I think it's a bit dangerous to have all this energy. And it's exciting, I don't know why. This thing has come closer to my body because I wanted it, but it's a bit overwhelming because it's so strong.
–The cloud is becoming clearer, it looks like gold. Like a golden haze. It's lost its darkness. It's like a Yang male energy that's coming in. [The client is shivering and sweating:] It's like an amazing struggle. Parts of me want it and parts of me don't. It's been a long, long time since I had it inside. I hate to think it's been away for so long. It has moved further and further away from me with time. It's not been part of me for so long. It has become unfamiliar. Now it's coming back, as if I had called it back. It feels SO REAL, it's incredible. It feels so right, so exciting. It looks like a white cloud with a golden edge. Before it was trying to creep in, but now it looks different, like a haze of energy above me. It's lost its scary feeling. It's a strong energy, and it's REALLY MINE. If I haven't had that in my body for thirty-five years, no wonder I was lacking energy.

In the weeks that followed this session, the client regained much of her self-confidence. The energy she recovered during the session stayed with her, and she was immediately able to display more authority in her activities. She went and told her boss what she thought of him, left her job and got a better one, etc. From a medical point of view, the psoriasis she had had for many years, and which had not responded to any treatment, started to improve.

Missing bits are much more delicate to deal with than entities. This agrees with a common principle of Chinese medicine that says it is easier to get rid of something in excess than to regain something that is missing. Most of my clients' entities are cleared after two to four weeks, i.e. two or three sessions. However, it usually takes much longer to get a client to become aware of a missing bit and to reconnect with it. The process involves certain practices akin to entity clearing, but also a lot of regression work with ISIS to explore the roots of the disconnection.

12.3 The Osiris process

Case study 12.3 Forty-two year old man.

–I have this picture of my chest, my physical chest. In the middle of it there is a big hole with irregular edges. On the other side of the hole there is space, a dark space full of stars. It's as if some parts of my heart were spread among the stars, like shining little pieces of glass, each linked to me by a cord. The hole in my chest is like a kind of gate, and on the other side there are the stars with all these parts of me.

Missing bits are not rare—we have all lost many of them. Cases 12.1 and 12.2 were extreme and gross. Having lost prominent parts of their astral body, the clients were unable to function properly and face the challenges of daily existence. However, we all suffer from similar missing bits, only they are more subtle. They do not relate to basic material survival but to more spiritual frequencies and functions. Their absence is insidious, and for the great majority of people remains undetected.

How did we come to lose all these astral bits? Some of them were dropped due to traumatic shocks of various kinds. Many others were simply forgotten by us somewhere in the astral space. They were not violently separated from us, we just never really bothered bringing them down to earth with us.

When you start reconnecting with these missing bits, you tend to realise that until then, you had never really been incarnated on the planet. You thought you were, but if one considers the totality of your being, you were hardly here. You were literally all over the space. The result was that you were sleeping your life instead of living it. Only when a gathering of all the parts has taken place inside your heart can you be fully present and find your real purpose on earth.

A classic way Zen Masters describe their enlightenment is: "When I walk, I walk. When I eat, I eat." This sounds very simple indeed, especially knowing it has usually taken them a few decades to reach that stage. However, after discussing the topic of the missing bits, this suddenly takes on a new meaning: "Whatever I do, I am fully here doing it." The totality of the being has been gathered, there are no more missing bits. A state of absolute presence has been achieved, and life has become a fullness of Self.

In John Borman's movie *Excalibur*, a beautiful retelling of the Arthurian legend, Parzifal brings the Grail to Arthur, who is on the brink of death. Arthur drinks from the cup, is suddenly revived, and exclaims: "I didn't know how empty my soul was until it was filled!" Then he goes on to win his ultimate battle. Arthur's *eureka* describes quite accurately the exalted enlightened state, the fullness of Spirit that comes from gathering all one's missing bits.

In the context of this book, this short chapter was designed to present very basic facts about missing bits, as one of the various mechanisms related to astral fragments. At a more advanced stage of the Clairvision techniques of inner alchemy, the topic is reintroduced on a higher mode. Just as certain practices are implemented to get rid of every single entity that could be latent as seeds in your astral body, others are used to find and explore every single cord that may be attached to you. The culmination of this process lies in the gathering of all the missing bits and their reintegration into your Self, just as the bits of the Egyptian solar god Osiris, who had been cut into pieces by Seth, were gathered and reintegrated—by Isis.

CHAPTER 13

POSSESSION AND EXTRAORDINARY ENTITIES

13.1 Borderlines

So far, nearly all the entities we have seen were nothing more than parasites. Their functioning was quite mechanical. Most of them were just astral fragments issued from dead people. The damage inflicted on the clients mostly came from the parasitic functioning of the entity, not from a deliberate intention to harm. If some were trying to hurt, like the revengeful Chinese brigand's head of case 10.1, it was only because it is the nature of fragments to repeat endlessly what has been imprinted in their substance. Thus their noxious character is just the continuation of the vicious emotions of the people they originated from.

Let me stress again that all of this operates quite mechanically. Due to the shattering of the astral body after death, a number of fragments are released, some of which happen to be toxic. There is nothing 'evil' or demoniacal about them, no destructive intention of some kind of dark force—just the automatic functioning of certain laws of nature.

From the point of view of the clearer the fragments and various other entities that we have seen so far are insignificant. If one has received the proper training, removing them is about as easy as taking a pebble out of a bucket. Once the client has observed the entity for a few days, as explained in Chapter 14, the clearing doesn't take more than a few minutes. Then the entity is recycled in such a way that it will never invade anyone else. Altogether, if a qualified clearer is available, having an entity is preferable to having a sore back: the treatment is painless and quick, and once cleared the entity never comes back. For this reason, I prefer to use the word 'clearing' rather than exorcism, as the latter is shrouded in a folklore that is totally irrelevant in the context of this work. There is nothing demoniacal at all about these insignificant parasites. In ninety-nine percent of cases an entity could never be called evil. It is just an energy in the wrong place, like the bacteria that are beautifully in place in a compost heap but

Possession and Extraordinary Entities

totally undesirable in your small intestine. They are exactly the same bacteria. If they are fine in the compost heap, why should we call them evil when they are in your intestines? In this thirteenth chapter, however, the time has come to take a look at the remaining one percent — those entities that can't easily be labelled 'just parasites'.

Case study 13.1 'A Tapeworm from Warsaw to Sydney'. Thirty-one year old woman, born in Australia of Polish parents. She works in the film industry. She suffered from the 'missing bit syndrome' described in the last chapter: lack of energy, lack of motivation, never feeling fully present in her daily activities, etc.

What are you feeling? –There is something coming out of my heart. It's like a growth, or like a scab. It wants to come out. It's attached to my heart, but it's foreign. It might even be a parasite. Before, it was right inside my heart. But now it's trying to come out.

What does it want? –It wants blood. It consumes it and it grows. It likes melancholic emotions, sadness, depression and suppression. It wants me to suppress my emotions, because it grows out of suppressed emotions. Inside my heart it looks like a tapeworm. The growth that comes out into the blood is like a waste product. It makes me never be present to what I am doing, always be in the clouds. It's trying to spread. It does not only send a waste product into my blood, it also sends baby tapeworms. I have this vision of myself being infested, it's awful.

Could there be a part of yourself that enjoys that? –NO!

You mean there is no part that enjoys being melancholic? –Oh, yes. There is a sort of comfort in being melancholic. It's like conditioning. There is a kind of resignation: you don't have to worry about anything. The worm helps that, it likes it. The worm is taking my energy supply. It takes my life, and my love. It drains. It sucks everything it can. It makes me unsure of who I am, of my boundaries. It puts me in a very undefined state.

Could it be that there are some foods you eat that it enjoys? –No. It's connected to the bloodstream. It gets everything it needs from my blood. There is something cultural about it. I got it from my mother, and she gave it to my sisters too. I can see it in their bloodstream. Their enthusiasm for life has gone. There is never anything fantastic for them.

[Spinning backwards in time with the ISIS technique.]

–It feels like seeing my ancestors in Poland, in a medieval castle. There is a woman talking to three knights, and there is a glass container. Someone gave it to her. The worm is inside it. It's in a liquid. It's alive. The vessel is

sealed. This woman is going to be beheaded, and she can't look after the vessel anymore. It's almost like she is the protector of the worm. After she is beheaded there is a fight in the castle. The vessel ends up on the floor, close to an open fireplace. The vessel is broken. And the worm breaks free. It's alive. It goes into the ground, and somehow it grows there, and finds its way to some people. When people get the worm they get sick at first but then they recover, they're OK. But they don't live any more, they just exist. The fire is lost. This worm is not only in me, it's in lots of people.
You mean there are other worms like yours, in other people? –No, it's the same worm, in all of them at the same time. It's spreading slowly to more and more people, not only Polish, Australians too.

In anthroposophical medicine, the blood is regarded as the physical vehicle which carries the life of the Ego. As Goethe puts it in Faust, the blood is more than just a physical substance, it is *"ein ganz besonderer Saft"*, a very special juice. In particular, it has a very special etheric layer attached to it.

The blood's etheric energy is extremely protected, but if an etheric parasite happens to invade it, it can create a state where the Ego is more or less disconnected from physical existence. This results in symptoms similar to those we described in the 'missing bit syndrome'. The client summarises this quite well when she says that nothing is ever fantastic for her sisters. Other signs, such as being in an undefined state, unsure of who she is and of her boundaries, are also clear indications that the Ego of the client can't express itself properly in her life.

The usual clearing worked wonders on this young woman. The presence of the worm could no longer be felt in her blood, and she quickly regained energy and enthusiasm. But if her perception was right – and from what I saw, that could well have been the case – this did not solve the problem for all the other people in whom the worm had proliferated.

13.2 Black magic and possession

Case study 13.2 'Indonesian Divorce'. Forty-two year old woman, receptionist. She is an Australian who married an Indonesian man and followed him back to his country. She was very badly received by the man's family. After a few weeks she became so depressed that she had to return to Australia alone. She suffered from constant anxiety, and a pain appeared in the right side of her chest. The feeling in her chest was so awful that she couldn't stop thinking of cancer, even though repeated medical investigations could not find anything wrong with her. Her depression was getting worse, she was more and more agitated, and couldn't sleep at night. Then she was warned by one of her in-laws that the Indonesian branch of the family "had hired a sorcerer to get rid of her". At first she found the information fanciful, until the very same was said to her by another in-law. The following is what she discovered through the ISIS technique.

–It feels cold and empty and frightening. It's like a black hole on the right side of my chest. The black hole came to me when I was in Indonesia, a month after I was there. I was just frightened. It was so different, and my mother-in-law really hated me.

–It looks like a pit, and inside it there is this person, made of mud. It's got a body but it's round, as if made of mud that has been roughly thrown together. Its head, body and arms are round. It's got eyes but it doesn't have a mouth, nor fingers, nor toes. It's jumping around, standing in my way. It doesn't want me to get past it. It cringes but it also jumps and laughs at me. It's like a barricade. This thing feels evil. The more I see it, the more it's obvious that the pain in the chest comes from it. It looks like a puppet made of mud.

How big is the puppet? –It's little. But I'm little too. It's smaller than me but it's very strong. It wants me to die. It hates me and I don't know why.

Where would it go if it left you? –To oblivion. Into space, into the blackness. But it doesn't want to. It's growing there, it's becoming part of me. It doesn't feel like me; it feels like something foreign, but it's trying to become part of me. Now it's not happy because I know that it's there. It's ugly and serious, and it laughs. I see it like a black hole I can go into, like mud growing in me, with this black person in it.

Are there foods that it enjoys? –No. It feeds on people's emotions. Especially sad ones, and hurt. Anything that can destroy.

From my perception, this entity could well have been projected onto the client by the conscious malevolence of someone hired by the in-laws. From Sydney, people often travel to Bali, Fiji, or Indonesia, for business or holiday trips. I have seen a number of clients who had just come back from these areas with various psychological and physical complaints, and who reported that a magician had been hired to harm them. In many of these cases, I have found and cleared similar noxious little beings that were playing havoc inside them.

Entities

At first I was very sceptical of any claim of sorcery. However, even though the clients described them in quite different ways, I found great similarities between these particular entities. While clearing them, I observed them carefully. My conclusion was that they were all made in more or less the same way — not an especially clever one, but quite efficient, and carrying certain easily recognisable signs. This led me to conclude that it was likely these entities had been consciously sent into the clients by local individuals with some occult knowledge. The entity in this case study certainly presented characteristic signs indicating it had been projected by one of these so-called sorcerers.

The clearing did not present any problem, and the woman's recovery was quick and complete. With a good technique, these black magic entities are not a problem to get rid of, no more than any other entities presented so far. But unless the victims have the good fortune to meet a qualified clearer their situation can become tragic. These little entities are particularly tenacious and unresponsive to any conventional form of therapy.

Let us now take a look at the misfortunes of an apprentice sorcerer.

Case study 13.3 'Face-ache'. Fifty year old woman, music teacher. Five years ago, after more than fifteen years of marriage, her husband left her for another woman. Then she tried to attack her rival psychically. Said the client: "I used to think of her in my mind and chop her head with an axe and cut her into pieces, and to call her 'face-ache' [a detail that will have an unexpected significance later on]. I had a book on witchcraft and I put a spell on her. I used to see myself as a skeleton, in a black cloak, chopping her into pieces."

Interestingly, the client looked neither like a witch nor an occult freak. She was a rather distinguished lady, the respectable mother of four children in their late teens, with an excellent reputation as a teacher. Whoever met her would have found it difficult to imagine her reading a witchcraft manual. Anyhow she wasn't very successful in her endeavours. Neither her husband nor his girlfriend were hurt, but afterwards she herself suddenly developed violent toothache. This was four years before she came to consult me. As soon as the tooth was fixed by a dentist, the pain turned into a horrendous neuralgia on the right side of the head. After that she suffered from unbearable pains that could be stopped only temporarily by major pain killers. She always heard a ringing in her head before the headache started.

What are you feeling? –A black cloud on the right side of my head. I can see a big yellow spider in it. When I look at the spider I hear a ringing in

Possession and Extraordinary Entities

my head, like each time I'm going to have a headache. The spider is lying on my head, on the right side. The whole thing looks quite ominous, not friendly.
Have you seen that spider before? –No, never. Although that's what it feels like sometimes, when I have the face-ache. Now the spider is saying "I want to get you."
Why? –It says "Just because I'd like to". It's revolting. But I'm used to it. I just know it's there and I have to live with it. And... there is a little ball in my jaw, about half an inch in diameter. It's one of the places where I have the pain. It's an egg. That's it, there is a baby spider in the egg. This one is blue, not black. And there is another big blue spider, another one. Its body is on my cheek bone.
And that one, what does it want? –The same. It wants to get me, to kill me. These spiders bite me. They're putting poison in my head and that's what hurts. Just after they bite me my mind goes into neutral and the pain starts. If I have alcohol, they bite me immediately.
Apart from alcohol, what do they like? –Anything negative. They feed on my negativity.

Now what happened to this client? I wonder. She did not come back for her next appointment. This was surprising, for she had spent four years unsuccessfully trying every possible treatment, and for the first time she had the feeling that she was getting to the root of her problem. Yet I never heard from her again. Entity clients have a high rate of cancellation and often find the most peculiar pretexts to arrive late for their appointments.

Case study 13.4 'Fatal Strawberries'. Twenty-six year old woman. Two years before, when visiting Lebanon, she was propositioned by a weird man, who was allegedly involved in magic practices. He offered her some strawberries, and insisted that she eat them in front of him. He didn't touch the strawberries. She ended up having sex with the man. Then she went back to Australia and never saw him again.

In the following months she started having strange dreams. She could also hear voices and feel presences in her bedroom, where she slept alone. The presences presented themselves to her as some kind of spirit guides wanting to instruct her in occult wisdom and mysteries. More and more often, she heard sounds of objects being moved around her during the night. She frequently woke up in the morning with bruises on her body without any reason.

Can you feel the presence at the moment? —Yes. It's advising me to make some changes in my life. To change my job.
What sort of job would it like you to have? —Child care.
What does it want? —It's watching over me and teaching me.
Yes, but if you tune into it, what does it want? —It's very faint. It's speaking in Lebanese and it says it wants me to learn about wisdom and to teach what I have learnt. And to teach other people to trust.
Does it feel more like a male or a female presence? —Male. Two males.
Are there foods that it enjoys? —What they like is my strength.
What do they get out of your strength? —They say "Trust us". I think that by strength, they are referring to sexual energy. They fulfil their own sexual interests.
What do they want? —They are trying to tell me that I should be promiscuous. There is also another presence, but it's not communicating.
What would they gain, if you were to be promiscuous? —That's where their interests are. They enjoy it when I have sexual intercourse. I feel they are watching me all the time. They're watching my breasts, and my sexual organ. It's like I can feel their pressure on my breasts. They're touching my breasts. And they're touching my sexual organ. They want to arouse me. And they like to watch my aura, my colours, and my spirit. They like the energy when I'm aroused.
Can they arouse you? —Yes. I just feel I'm being touched. It's physical and they say that it's my fault, because of the way I sleep with my legs apart. Last year, I often had the feeling that I was being pinched during my sleep, and then I woke up with bruises on my hips. The feeling of presence was exactly the same. Once I was very drunk and it was like having sexual intercourse with them during my sleep. I felt I was being punched and beaten, and I woke up having an orgasm. They both want to have intercourse with me. Since that night it's happened quite often that I've had sex with them at night and that I've woken up having an orgasm.

The concept of entities who have sexual intercourse with human beings is far from new. It was already widely discussed in Europe during the Middle Ages, where the Latin names *succubi* and *incubi* were used to describe them.

Incubus comes from the Latin *incubare*, to lie on. An *incubus* was understood to be a male entity that comes to have sexual intercourse with women during their sleep.

Possession and Extraordinary Entities

Succubus comes from the Latin *sub*, under, and *cubare*, to lie. (The English verb 'to succumb' comes from the same root.) A *succubus* is a female spirit, that comes to have sex with men during their sleep, sometimes causing wet dreams.

In a Chinese Taoist treatise on sexuality, the *Su Nu Jing*, similar entities are mentioned. At the end of its first book, the *Su Nu Jing* describes how, either due to lack of a partner or to an excessive temperament, some people practise coitus with a ghost. The treatise adds that the pleasure from such coitus is much more intense and addictive than that of the normal sexual act. The *Su Nu Jing* therefore regards this as a disease and suggests that if not cured, the person will die after a few years. Interestingly, the treatment indicated by the treatise is as follows: the patient must practise uninterrupted coitus for several days without ejaculation. After implementing this therapy for a maximum of seven days, the cure is guaranteed.

This very respectable Taoist scripture then tells how sceptics can convince themselves of the reality of this disease: Let them go and wander in the mountains or in a forest, in spring or autumn, thinking of absolutely nothing but coitus. After three days of this uninterrupted practice, they will experience alternating fever and coldness, anxiety, and visions (men see women, women see men). The pleasure of coitus with such visions is much more intense than with human beings. The text then concludes by saying that this disease is very difficult to cure—but doesn't specify whether the sceptics will have caught it or not after their three day experiment.

In Wagner's opera Tannhäuser, one finds a similar story. Tannhäuser is having a love affair with a non-physical entity, in this case Venus herself. The opera narrates all the difficulties Tannhäuser experiences in freeing himself from the enticing attraction of Venus' cave. Wagner's treatment, however, is radically different from the Taoists' — he advocates prayers and penance.

13.3 Pseudo spirit guides

In case study 13.4, there was an important detail; the entities were trying to appear to the client as her spirit-guides. Let us spend more time analysing this treacherous mechanism.

Case study 13.5 'Take me to India'. Forty-two year old woman, public servant, married, with three children. Lately she has been in touch with a 'spirit-guide' that sends her information and guidance about herself, her future, and life in general. The guide wants her to give up her job and her family to go and live in India, so that her spiritual Self can blossom. She is very excited about this guide, she 'channels' him all the time.

A friend of hers, who was not convinced by all these visions, suggested she come to discuss them with me. Here is what appeared when we looked at the guide in an ISIS session.

Why exactly is he with you? —He doesn't want to reincarnate. He doesn't need to. Also he has a fear of the suffering life brings. And he enjoys being around me.

Does he like it when you have sexual intercourse? —Yes. He gets energy from it. He watches. He is with me all the time. He says he is not a bad fellow. He cares about me and he wants to help me. He says he likes me sexually too. He doesn't like me to be with my husband, he wants to keep me all for himself.

Have you ever made love with this guide? —It's happened a few times. But he says it's not to happen any more. I should just remember it.

Are there some foods that it enjoys? —Chocolate.

Does he sometimes push you to eat chocolate? —Now and then. But now he wants me to have a better diet.

So what does he want? —He's going to stay with me, because he cares about me. He says I need his protection. He says the time will come when I have fulfilled what I had to do with him, but for the moment I still need him.

By the way, where does he get his energy from? —From me. From my heart.

Let's summarise. You have a spirit guide that pushes you to eat chocolate and that watches you when you have sex. He doesn't want you to be with your husband because he is jealous, and he wants you to masturbate while thinking of him. Moreover, he takes his energy from your heart. Isn't that a bit suspect, altogether? —Yes, maybe.

Is it nice to have his presence around you? —I really like it. It makes me feel much stronger. I feel I can just live my life without doing the wrong things. He helps me to make decisions, because I can't make them for myself. He thinks you've got the wrong idea about him. He thinks that he's OK. You're giving him a hard time.

This chapter is entitled 'extraordinary entities'. Unfortunately, I am afraid there is nothing extraordinary at all in mistaking an entity for a spirit guide. This has become only too common these days, when 'talking to one's spirit guides' is increasingly regarded as a sign of status, just as

Possession and Extraordinary Entities

credit cards or the membership of certain clubs used to be. My perception is that many of the people who presently think they are in contact with a spirit guide are in reality in contact with nothing more than an entity.

In most cases, for instance in the example above, the so-called guide is a fragment, an astral bit which originated from the shattering of the astral body of a dead person. Then the 'guide' parasitises the client in exactly the same way as did all the fragments we saw earlier. It drains life force, creates food cravings and sexual fantasies, adds its own emotions on top of those of the client.

The client from our last example did not like what had been revealed about her spirit guide. She never came back.

Let us look at another similar example.

Case study 13.6 'The Black Nun'. Twenty-six year old woman, nurse. She had been smoking more than twenty cigarettes a day for about fifteen years. She had ended a relationship three years before this session and since then had been unable to start a new one. She felt blocked, as if she was building up barriers against anybody trying to approach her. During the first minutes of the ISIS session, the young woman seemed to be quite embarrassed by what she saw. Then she declared:

–There is a nun inside! I can see her. There is something mysterious because she won't show her face.
What does she look like? –Dark. I can see the black and white habit clearly. But there is a dark space instead of the face. She's bowing her head down. At the moment she looks a bit uncomfortable, because we are looking at her.
Have you ever seen her before? –Yes. I saw her a few times in meditation. But I thought she was around me, I didn't know she was inside. She's my spirit-guide. I've often been aware of her serene feeling around me. I like her. I feel good with her. I like working with her.
Does she like it, when you work as a nurse? –Yes. Very much. She helped me to decide to study nursing. It made her very happy when the decision was taken.
If you tune into her, how do you feel? –I don't like to show my face. I like very much to be alone.
Who likes to be alone, you or her? –Me. Or maybe it's her. Or maybe it's because I'm like this that I can relate to her. Or... I think it's her. But she likes to work with me on people. She likes to do as much as she can for them. Whatever is possible, because she has great love for them.

How does she react when there are men around you? —She does not like me being involved because she believes in purity. But maybe, in reality she is afraid. Yes. That's it. She fears men.
Do you sometimes feel the same? —Oh, yes!
In reality who is afraid, you or her? —Mm! [Confusion] Her. I think she is the barrier. I think she manipulates them away.
Which area of your body is she most connected with? —The heart. She lives there. That's why she won't show her face. Because she feels guilty for being there and for blocking everything. She is terrified of the world, so she hides inside me.
Does she ever get bored, inside? —Yes, lots.
Why does she stay, then? —Because she feels safe there. She says she just wants to send you love, she is not asking for anything more.

This last answer sounded like a flagrant lie, both to the client and to me. The client's reaction after this session was quite different from the one we saw in the last case study. She suddenly realised that she had been tricked, and that the 'inner voice' she used to trust was not her spirit-guide, but something much less noble living inside her like a parasite. The shroud of serenity and holiness around the nun was now seen as a cynical masquerade, the source of the barrier she felt between her and the world.

After she watched this fragment for the usual ten to fifteen days, I cleared it. Interestingly, while I was performing the clearing, she could suddenly see the nun's face. She described it as "a skull under the hood, putting her hands over her ears so as not to hear the sounds of the clearing."

Unexpectedly, within the ten days that followed the clearing, the young woman completely gave up smoking. Her way of relating to men gradually changed, and she started reopening to the world.

13.4 Dark forces

Case study 13.7 'High Spirits in Thailand'. Twenty-nine year old man, engineer. He had never done any meditation or any work on himself. He had never taken any drugs, never had any psychiatric problems. He had always been a rational and scientifically oriented type of person, grounded and full of common sense.
He went to Thailand for a holiday and, while visiting a grand palace in Bangkok, experienced "the weirdest feeling of my life". Even though he

Possession and Extraordinary Entities

couldn't define exactly what was happening, he suddenly felt transported into a completely new dimension of himself. The following night, at his hotel, he had his first out of the body experience. Then, said the client: "An evil spirit came to me and promised it would give me eternal life. But I could see that in reality it was eternal death." The client reported that the spirit took him travelling in space, seeing events of the future and all of his next life, visiting heaven and other worlds, and being greeted by the gods. Everything kept on unfolding extraordinarily fast, creating a feeling of superconsciousness. For days the man felt utterly euphoric and agitated, unable to sleep at night. He repeatedly had the impression that he was levitating over his bed.

This manic state persisted after he returned to Australia. He could read people's minds, know in advance who was going to ring or open the door, and see non-physical beings. Unable to cope with the intensity, he even tried to commit suicide twice.

As a medical practitioner, I have dealt with several schizophrenic patients, but this man did not feel like one of them. Moreover there was an intensely concentrated aura of energy around him when he arrived at my place.

Can you see it? –It's black. Like an oval ball. And it has an obscure kind of face. If anything it would look like a panther.
To which part of your body is it most related? –Around my navel. It just seems to be watching me, trying to reassure me. It feels blacker than just black. It seems to be watching me, waiting for the right time to possess my Spirit. It's like a huge black cat. It's not unfriendly, but it wants my Spirit.
What would it do with your Spirit? –Control it. Possess it with its own strength.
Could it really do that? –Maybe it could. Or maybe it couldn't, but it makes you believe that it can. It gives you the feeling that it can make you die, that it can make life slip out of your body, and then it can take over and you belong to it forever. I had the same feeling in Bangkok. There is an incredible life force concentrated in that cat. It creates a very euphoric feeling for me. My body feels completely hyper, but cool at the same time. I can go for weeks without sleeping. No stress, no fatigue, just power, raw power. It makes you feel fantastic and you can accomplish heaps. Every single thing you do succeeds.

The second part of the story was much less exciting than the first. I cleared the cat, and everything stopped. The client became a rational, grounded person again—perhaps with more interest in supernatural phenomena. The fact that clearing the entity stopped the flow of non-physical perception was also an indicator that the client was not a schizophrenic. With schizophrenics, a clearing is unfortunately not enough to restore balance.

That entity was not a fragment.

Case study 13.8 'Follow Me'. Forty year old woman.

–It's like an overwhelming, very authoritarian type of presence. It wants to put me down. Never happy. It says that I'm worthless, that I'm nothing more than a spoiled brat and that I have no right to anything. It drags me down into the dirt.

What does it gain out of it? –It makes it right. If I'm down then it can go up a notch. If I was completely down, it could take over, it would get stronger and spread. Into lots of people.

What does it look like? –It's dark. It looks like a human shape, a human face. It has a head cover, like a kind of helmet. The face is in front of me, and behind it there is a light. Before it was in front of me, the face was in that light. It came out of it a few weeks ago.

If you tune into that light, what can you feel? –That's really bad. It's like a gold light, but not healthy, and there is black around it. This light is wrong. It's calling me. [Tears:] It wants me to walk into it.

What would you gain, if you walked into that light? –I'd stop being miserable, I'd get out of the dirt. I would start shining that light. It would support me and make me very powerful. It's awful, there is something wrong and twisted about it, something evil. And it's incredibly powerful. It's like a way. It's not just after me. It would like you to go into it. It's after you, it wants to wreck your school.

How would it do that? –If you, or just one of the people around you went into it. It's like a fungus, it spreads. It's something that can turn people into poison. But it can make them so powerful that they would hesitate before saying no.

Once more, the end of that story was quite common — I cleared the face and the pernicious light, and everything stopped. But this entity was not a fragment either. After it had been cleared, the client commented: "That was the scariest thing I have ever seen."

Possession and Extraordinary Entities

The topic of extraordinary entities is vast and colourful, and one could write entire books about it. However, there is something biased in focussing on it too much. As the reader can judge by having an overall picture of the case studies presented in this book, the vast majority of entities are just fragments, i.e. simple parasites that react mechanically and don't have anything evil about them. By overemphasising a few dazzling cases, one tends to create false ideas. All the cliches and folklore surrounding exorcism are responsible for distracting serious investigators from getting interested in an important topic. The knowledge related to entities can relieve so much suffering that one can only regret that so few people presently know how to perform a proper clearing.

CHAPTER 14

EXPLORING AN ENTITY

14.1 *An entity is dealt with in 3 stages*

In the Clairvision School's style of work, an entity is dealt with in three stages:

Firstly, discovery. The client becomes aware of the entity during one or more ISIS sessions. The space is open for the client to see and feel the entity, to explore what it wants, and where it may have come from.

Secondly, observation. For a duration of about ten to fifteen days, sometimes more, the client keeps on watching his/her entity during his/her daily activities. The aim is to discern the action of the entity, which up till then had remained unseen. Clients are encouraged to carefully observe their emotions, thoughts and reactions to their environment, in order to differentiate between what comes from the entity and what comes from various parts of themselves.

Thirdly, comes the clearing. Taking the entity out of the client and sending it into a very special frequency of light, in which it will be processed and recycled.

The purpose of this chapter is to explain the reasons and methods of the first two phases, discovery and observation.

14.2 *A clear warning*

A clear warning must be given before starting: dealing with entities requires highly technical knowledge. Becoming an entity clearer takes a lot of time and effort. It is only if the technique is impeccable that the process is safe, both for oneself and for the client. A number of pitfalls, some of which will be outlined in the next chapter, can be encountered and can lead to catastrophe. I therefore strongly advise readers *not* to try to explore or deal with entities, unless they have received a special training and initiation to do so.

Exploring an Entity

At the Clairvision School we offer powerful techniques and encourage students to practise them as much as they can. Thus the school has given the knowledge of the intense ISIS technique of regression to hundreds of people, and let them use it freely. However, when this book was written, only a handful of people had been taught how to clear entities. Let me repeat again that **clearing entities is potentially dangerous and should therefore only be performed by practitioners who have received a complete training on how to do so.** The purpose of this chapter and the following one is *not* to teach you how to explore and clear entities, but to gain a better understanding of them by seeing certain aspects of the way in which they are processed.

Sometimes the question is asked: why bother exploring the entity, why not just have the clearer take care of it and send it away into the light, and forget about it. The general perspective of the Clairvision techniques is to enable people to know themselves. Whenever a problem arises, it is for a particular reason, and there is something to learn from it. One certainly learns a lot by exploring an entity. Firstly, one learns about entities themselves and the space they come from, which can only enhance broad-mindedness. Secondly, entities do not arrive by chance. If they get attached to you it is because something in you has attracted them, and because a breach in your defence system has allowed them to penetrate. Having the entity cleared without exploring it would be like refusing to look at the weak point which allowed the entity in. It would be a way of disowning the problem. On the other hand, by knowing exactly what happened, one can work at correcting the imbalance, and save oneself further problems.

14.3 The discovery phase

This discovery phase mainly takes place through the ISIS technique, as was discussed briefly in the first chapter.

The first important principle is never to tell clients that they have an entity. It is far preferable to take them into the ISIS state and get them to see it for themselves. There are a few obvious reasons for this approach.

First is a desire for objectivity. Obviously, any inner work deals with subjectivity. Even so, the more objective one can remain while exploring this subjectivity, the better. Entities are not fluid and changing experiences like dreams, they are 'things' made of etheric and astral matter. Clients are encouraged to stick to what they see and feel, and not to make up anything. The ISIS techniques never use hypnosis, suggestion, positive affirmations or creative visualisations. They operate an opening of perception through a direct activation of the body of energy. They are aimed at seeing the world the way it is, not the way we would like it to be. As long as clients do not see the entity or feel its presence, I never mention anything about it. It is only when both they and I have perceived 'the thing' that we say there is an entity.

There is another good reason why it is far preferable not to tell people they have an entity. Suppose I tell a client: "I can see this black octopus-looking entity sticking to your back. It's got tentacles around your neck. It's sucking your life force and it wants to make you die a slow and painful death. But don't worry, everything is OK, it's just an entity." How do you think the client would react? Yet I regularly see people who have been told similar things by a psychic during an aura reading.

On the other hand when clients discover an entity by themselves, experience has shown me that they cope remarkably well. The ISIS state creates an inner opening, a closeness to the Self. It puts clients in touch with deep soul forces, so they keep a certain equanimity whatever they may see or feel. When making contact with an entity, they immediately realise that it has been there a long time. The fact of seeing it there is not in itself going to make things worse. Rather, it will allow them to understand what is happening inside them, and take measures accordingly.

Another important point is that even if clients are not consciously aware of the presence of their entity, on a subconscious level they know perfectly well it is there. ISIS puts them in touch with the part of themselves that knows. Therefore everything unfolds quite simply and naturally for them. One does not have to do anything to reassure them. One just has to let their own knowingness and wisdom operate for them to see they are safe. After all, it's just an entity.

In ISIS, I have seen many sensitive or strictly materialistic clients describe mind-boggling entity stories, full of horrendous details, and yet remain quiet and unaffected. In years of practice, not one client has panicked or asked to interrupt the session because they could not cope with the entity they were discovering. Had they seen something similar on TV, they would probably have reacted violently. But in the purer consciousness and the knowingness of the ISIS state, they were able to accept what they saw about themselves without really being affected.

Another principle in this work is that the clearers never say how they see the entity as long as the clearing has not been completed. Otherwise they would pollute their clients' spontaneity with their own projections. It is a form of respect for clients to let them describe what they feel and see with their own words. When everything is finished, impressions can be shared freely. However, in most cases what clients discover themselves is so precise and clear that they do not even think of asking the clearer's opinion as to what 'the thing' may be or what it looks like.

14.4 Getting in touch with the entity

During the first minutes of the session, the clients are connected with the ISIS state. The method implemented for this purpose has been described in the practical part of my book *Regression, Past-Life Therapy for Here and Now Freedom*. It consists of activating the client's third eye and getting in touch with the inner space of consciousness.

Once this is achieved, the next step is to make contact with the entity. In certain cases this happens instantaneously. The entity is nested conspicuously in the client's energy, so as soon as one looks inside it appears automatically. In many other cases, the entity will be discovered fortuitously, for instance while undergoing a regression process.

If an entity is suspected but the client remains unaware of it, the following method can be used to get in touch with it. Suppose the entity is in the chest, the client is told:

"Become aware of your chest. Feel your chest from the space in between your eyebrows."

After one minute, the next instruction is:

"Compare the right and left sides of your chest. Which side feels heavier?"

After the client has answered, the next question is:

"Which side feels darker, the right or the left?"

Then,

"Which side feels thicker [or denser], the right or the left?"

The order in which the three questions (darker, thicker, heavier) are asked does not really matter. The same method can be implemented on any body area.

In the ISIS state, when clients have an entity in a certain part of their body, in the majority of cases they perceive that part as heavier, thicker and darker. However, this is not absolute. In certain cases, one of the three characteristics is missing. For instance the client may feel the area is heavier and thicker, but not darker. In other cases, it may even be that one of the characteristics is inverted. Thus the client may perceive the area is heavier and thicker but brighter than the other side. Or the perception may be a darkness that is lacking substance and does not have as much energy

Entities

as the other side, and therefore feels lighter. Fundamentally, entities are something added to the client's own energy. So the most common finding is that the area where the entity is located feels heavier and thicker.

During the very first minutes of the session, it is not uncommon for clients to experience confusion as to which side the entity is on. For instance, they first feel it on the left side of their chest, and then change their mind and decide it's on the right. Then, in most cases the location remains fixed till the end of the clearing process.

The next instruction is:

"Remain in the space in the middle of the eyebrows, and from there tune into this area that's darker, denser and thicker" (or whatever adjectives were used by the clients themselves).

Then:

"How big does it feel?"

At this point it will happen that the client has already made full contact with the entity, feeling its presence and seeing what it looks like. In that case one proceeds with the other questions on the list. If the client only feels a blurry dark cloud, that too is often enough to proceed, and the entity is gradually perceived more clearly as the following questions are asked. In other cases, however, the client does not feel anything more than a 'darker, thicker, heavier' area and a few more minutes are needed to get in touch with the entity.

It is not difficult to find out whether the client has made contact or not. If not, he or she usually finds it impossible to answer the following questions, or the answers are vague and incoherent.

If the client cannot find any difference between each side of the body in this particular area and cannot relate to any of the other questions, it is preferable to give up for the time being. Use the ISIS session to explore some more tangible issue, and come back to the possible entity later on. With this method, it is only if clients themselves perceive the entity that one should accept there is one.

14.5 *Exploring the entity*

The purposes of the exploration phase are:
- to get to know the entity, what it looks like, what it wants;
- to realise what interference it has created in the client's life;
- to look at the part in the client that benefits from the entity's presence, and may even want to keep it;

• to find out, whenever possible, when the entity was first attached to the client, what circumstances allowed it to break in, and where it came from;

• to prepare for the next phase, during which the client will have to observe the entity during daily activities and be aware of any interference coming from it.

We will now look at a list of questions which can be used in order to achieve these purposes.

Having realised the presence of a darker and heavier area while in the ISIS state, a good way of continuing the exploration can be:

"What emotions or feelings could be related to this darker, heavier (or other adjectives used by the client) **area?"**

Clients can usually relate easily to this question, which helps them get more in touch with the entity.

"Does it have a shape?"

At first, the client often sees his entity as a blurry dark cloud. The full shape is only perceived after a few more questions. That is why this question is formulated in a somewhat vague way. Only much later in the session should one ask: **"What does it look like?"**

It is not unusual for the very first image of the entity to be unclear and distorted. Then, a few seconds later, a new image is perceived which will usually remain unchanged until the end of the clearing.

"Have you ever seen it before?"

"Can you feel something like a presence attached to it?"

Some clients find it difficult to understand what 'presence' means. So it is better not to insist too much if they can't answer. One can come back to this question later, when more aspects of the situation have been seen.

"Does it feel like something foreign or like a part of yourself?"

If the client seems to hesitate, one specifies:

"Don't think, just say what you feel."

Certain clients have a mental background that doesn't allow them to consider that anything inside them may be something other than a part of themselves. While in the ISIS state, it is essential for clients to stick to what they perceive instead of going into their mind and thinking of a clever answer. For the same reason, in ISIS whenever a client seems to hesitate about a question, it is preferable not to insist. Otherwise some clients will make up meaningless answers.

"What does it want?"

This is a key question, one that will have to be repeated several times during the session.

When referring to the entity, it is preferable to use the word the clients themselves have used to qualify it, for instance 'the cloud', 'the presence', or 'the thing'. In the following questions, I will use 'the thing', but it should be remembered that in practice it is the client's words that are used.

"Could it be that there are some foods you eat that 'the thing' enjoys?"

If yes, follow by:

"What happens to the 'thing' when you eat these foods?"

"Does it sometimes generate cravings for these particular foods?"

"Apart from ...(what the client said)**, what does it like?"**

"What happens to 'the thing' when you smoke?"

"What happens to 'the thing' when you drink?"

"Could it be that it sometimes pushes you to drink or to smoke?"

"Does it sometimes interfere with your sexual activities or desires?" If yes, **"How?"**

"What else does it like?"

This question, similar to "What does it want?", allows clients to discern the essential nature of the entity. It can be repeated more than once, because clients often discover new aspects of the entity's personality as they advance further in the session.

Whenever the client finds out something the entity seems to want, the following question is:

"What happens to 'the thing' when you have/do this?"

Another similar question that can be used in order to explore the entity's motivation is:

Exploring an Entity

"Among your activities and the various things you do, are there some that 'the thing' enjoys?"

Followed by:

"What happens to 'the thing' when you do that?"

One can also ask:

"Are there some other things it wants you to do?"

Or:

"Does it dislike some of the things you do?"

Depending on the situation, one may complement these questions with others such as "**Does it like or dislike some people around you?**", "**How does it react to your husband/wife/boyfriend/girlfriend?**", "**How does it react to your children?**", etc.

And also, sometimes:

"Does it like you to wear certain clothes or any colour in particular?"

"What does it gain out of being inside you?"

Or:

"What does it gain from you?"

Or:

"How does it live? Where does it take its energy from?"

Some clients find it easier to understand what 'presence' means after answering the important question:

"Do you sometimes get the feeling that 'the thing' is watching you?"

"Does it sometimes make voices in your head?"

If yes:

"What do the voices say?"

A revealing question can be:

"How does 'the thing' react while we are looking at it?"

This can be a good time to repeat the essential question:

"What does it want?"

Entities

"If 'the thing' could speak through you, what would it say?"

Another revealing question is:

"If 'the thing' could take you over, what would happen?"

Followed by:

"Does it sometimes happen that it takes you over?"

If yes:

"What happens then?"

"What does it look like?"

At this more advanced stage, as the client is now fully in touch with 'the thing', there is more chance of getting a detailed description. In the vast majority of cases, the aspect of the entity does not change significantly until the end of the clearing. During the phase of observation, it can change size slightly. It may get a little bigger, for instance if the client uses toxic foods or drugs. However these variations remain moderate, and a spider usually remains a spider, a head remains a head, etc. Apart from a few extraordinary cases, an entity that changes form all the time, being an octopus at one stage and then a little person or an insect later on, is quite likely to be a mental fantasy rather than a real entity.

"Does 'the thing' sometimes move, or does it always stay in the same place?"

If it moves, follow with:

"What happens to you when it moves?"

"Is the thing related to one of your organs?"

Many clients do not have the faintest idea or perception of where their organs are located. Whenever this is the case, the question is irrelevant, of course.

"Does 'the thing' rather feel male, female, or neither?"

Apart from particular cases, this question of gender is of little interest, because clients usually can't really relate to it and often give a different answer if it is repeated a few minutes later.

Among the key questions are:

Exploring an Entity

"Could there be parts of yourself that benefit from the presence of 'the thing'?"

"What kind of benefits do you gain from the presence of 'the thing'?"

The answers to these questions often reveal why that particular entity came to be attached to the client. Working on these issues would not be enough to clear the entity, but still it would help improve some of the client's major weak points.

Trying to find out when, how and why the entity came in is an important part of the exploration, even though in many cases it won't be possible to get more than vague indications. First, some simple questions can be used, such as:

"Does it feel like 'the thing' has been with you for a long time?"

"When did it come in?"

This question often proves insufficient, so the ISIS regression techniques are used, in particular the spinning techniques. A possibility is to get clients to spin back in time and re-experience what their body looked and felt like ten years before, to see if 'the thing' was already with them. Then, as long as the source has not been found, one keeps on moving backwards every five or ten years until the womb is reached. As the womb stage is a critical time to catch entities, one explores it in detail. Then, if necessary, one keeps on moving further back in time.

Once the time of invasion has been sourced, it is valuable to explore if some part of the client actively attracted the entity and the reason for this, as well as all other related circumstances.

Even if no source can be found, clients sometimes find it easy to answer the question:

"Where was 'the thing' before it was with you?"

This often triggers fascinating answers, by getting the client to tap from the memories inscribed in the entity.

One can conclude the exploration with:

"If you were to die, what would happen to 'the thing'?"

"If 'the thing' were to leave you, where would it go?"

"Would it rather go into the earth, or water, or wind, or fire, or space?"

The thinking behind this last question is that entities have an affinity with the element they are most related to — earthy entities aspire to return to the earth, and so on.

Sometimes, it may also be appropriate to ask the difficult question:

"Apart from you, is it also attached to somebody else at the same time?"

The order in which the questions are asked can of course be modified according to the individual characteristics of the entity and of the client. In ISIS, an important principle is to ask only questions that clients can answer. Otherwise, they have to use their usual mental consciousness in order to make up an answer, which projects them out of the expanded state of perception. In particular, at the beginning of the session clients are not in touch with their entity enough for elaborate questions to be asked. Therefore one first deals with elementary points, such as "How big does it feel?" or "Does it have a shape?" This gradually allows the client to make full contact with 'the thing'. Only then does it make sense to ask the more complicated questions on the list.

It is essential to understand that it would not make sense to ask any of these questions in a normal conversation or interview, without being in the inner space as contacted through the ISIS technique. It is only on the basis of an expansion of perception that such an exploration can be significant. Otherwise the best the client could do would be to guess, and the whole thing would turn into a mind-game, a useless masquerade.

Another essential point is that it is not enough to explore an entity to get rid of it. For this a proper clearing is needed. Conducting an exploration as mentioned above without following it with a full, regular clearing could even make the situation worse, by exacerbating the symptoms caused by the entity. It is fine to discover one has an astral spider in the neck and to observe it, as long as one knows for sure it will be cleared within one or two weeks (as is the case whenever a proper clearer is at hand). Otherwise, it can easily turn into a nightmare. I must repeat that only skilled 'not-born-yesterday' clearers should deal with entities, or else anything can happen.

14.6 Concluding the exploration session

After the ISIS session, and before giving the instructions related to the phase of observation, it can be helpful to explain a few points of theory to the client. Many have preconceived ideas about entities and exorcism. It is therefore necessary to de-dramatise the situation, by sharing with them some of the information that is developed in this book.

Exploring an Entity

In particular, one must emphasise that entities are nothing more than parasites and that there is nothing extraordinary or evil about them. It should be clear in clients' minds that they are not 'possessed', and that the process of clearing the entity will be simple and quick. Usually that does not cause any problems. In the simplicity of the ISIS state, clients are able to see for themselves what 'the thing' is really about. Therefore they rarely come out of the session worried. It may also be beneficial to emphasise that, however awkward or unpleasant it may sound, discovering an entity is rather a good sign. It will be cleared within a short time, thereby allowing one to progress and improve one's inner situation significantly. Remember it may take years of work to heal a trauma from early childhood; whereas when following this process, after a maximum of three weeks the great majority of entities will have been cleared. And they do not come back.

At the same time, one should be careful not to let clients use entities as scapegoats responsible for all their difficulties. For some people, entities are an easy way of disowning their problems. I have had clients who, whenever they found something they didn't like about themselves, asked me "Are you sure it couldn't be an entity?" Or some people, if they discover an entity that is related to some inglorious emotion of theirs, immediately say "Oh! Good! Then it is not me, it's the entity!" Such an attitude is not responsible, and not conducive to real progress. These clients must be reminded firstly that 'the thing' did not arrive by chance, and that if it came to be attached to them it is fundamentally because of some disposition they shared with it. Secondly, one does not get rid of psychological problems simply by having an entity cleared. The clearing will facilitate the work that must come next, but can in no way replace it.

14.7 The phase of observation

After the ISIS session in which the entity was discovered and explored, the client has to undergo a phase of observation. The following instructions are given:

First, do not try to push the entity away and do not ask it to go. The entity has been inside you for a long time and if it was enough to tell it to go away for it to do so, it would have departed long ago. It is one of the basic features of entities that they do not leave that easily.

Moreover, by asking the entity to go or by trying to force it out, one might cause it to hide, which could only complicate and slow down the process. From my experience if, one or two weeks after discovering an entity, a client comes back saying that it has disappeared, it can be for two reasons. Either 'the thing' was not an entity but a mental illusion of some kind, or it was a real entity—in which case, ninety-nine percent of the time it has not gone, it is hiding.

The best possible attitude towards the entity is neutrality. Sending it negative thoughts would in many cases reinforce it, for entities usually feed on the client's negativity. However, sending it love wouldn't necessarily help either, for this may also feed the entity and reinforce its emotional ties to the client. The advice given to clients is: Just be aware of it, just watch it, but do watch it. Remain aware of its presence as often as possible. Become like a zoologist observing the behaviour of some kind of creature. Do not hesitate to make notes, to analyse your feelings. If carefully done, it can be a most instructive experience—not only about 'the thing', but also about yourself. In particular, whenever you experience emotions, cravings, or desires, immediately go inside and try to find out whether it comes from the entity or from a part of you. Once you have made contact with the entity, this is usually not very difficult. It allows you to clearly discern what interference is caused by 'the thing'.

If a craving or an emotion is felt, the client is encouraged not to suppress it, but to look for its source. From the point of view of the process, the problem is not whether the chocolate will be eaten or not—it is to find out if the craving comes from the entity. The same principle applies to emotions and reactions of all kinds.

The work of observation should be extended to all aspects of the clients' lives — their occupation, the people they live with, the activities they favour during their leisure time, their health and physical symptoms, their eating habits, etc. It should lead them to know for sure:
- what the entity wants, what it likes and dislikes;
- how it influences their emotions, desires and choices;
- what inside them is themselves, and what is the entity.

Once this work is completed, the client is ready for the clearing.

CHAPTER 15

CLEARING ENTITIES

"The great light, which is the Heart of God..."
Jacob Boehme, *Aurora*

15.1 What problems can occur while clearing an entity?

Many. The first is that entities are tenacious creatures. They don't let themselves be cleared unless some special procedures are implemented. The idea that one could get rid of an entity just by telling it to go away, or by visualising it away, seems totally fanciful to me. If certain clearers appear to do so, it is because they are backed by a particular energy, a power that performs the clearing while they speak their words or do their visualisation. However, if anyone just attempts to tell an entity to move away and disappear, or tries to visualise it away, the entity is most unlikely to go. If it can't be seen or felt any more after the exercise, there is a ninety-nine percent chance that 'the thing' is hiding, not cleared.

A process that works on the psychological level only, trying to get the client to 'let go' of the entity, would be just as inappropriate. Before I knew how to clear entities, I remember doing up to twenty-five sessions with a client on a particular entity—not even a difficult one, just a little fragment. We explored again and again how and why the entity had come in. We found out and released various past-life traumas that had created the weakness for the entity to come in, and might have attracted it. The client reached a point of neutrality regarding 'the thing', neither liking nor disliking it. Yet the entity was still inside the client, just as it was the first day we looked at it, totally unresponsive to our efforts, and determined to hang on.

The reasons for this are obvious. Firstly, getting the client to let go of the entity is not enough to get the entity to let go of the client. The entity depends on the client for its survival. Asking an entity to let go is like saying to an animal "Please stop breathing and die." Secondly, even if the entity was ready to let go, it is intricately entangled in the etheric and astral bodies of the client. It is literally trapped inside the client's structure. A particular force is needed to pull it out of the client's energy. This must be done fully and properly, for a partial clearing is worse than no clearing at all: entities which manage to resist a clearing, even partly, often get rigidified and are harder to clear later on.

Furthermore, ejecting the entity from the client would not be enough. For where would the entity go, then? An entity that has just been evicted is quite contagious. It will do anything it can to attach itself to whoever is around. Maybe it will jump straight into the 'clearer' (that is usually what they try first). Or maybe it will hide under the sofa, waiting for the next client, or the clearer's next nap, to creep in. Maybe it will find its way into the children in the other room, or into the next-door neighbour.

This implies that the clearer, apart from having the power to extirpate the entity, will also have access to a (non-physical) safe place where the entity can be disposed of and processed in a way that renders it harmless.

Even if the clearer's technique is impeccable, at the moment when the entity is ejected it is not unusual for etheric and astral bits and pieces to be released. Most of them go straight into the clearer. It is imperative that the clearer has enough perception to recognise when this is happening, and knows how to eliminate them. Otherwise, these little fragments will accumulate, and in the long run they could be responsible for health deterioration and possibly severe diseases.

Apart from that, the clearer can receive threats from entities. You may know certain dogs who are utterly sweet and domesticated and would never bite anyone. One day the dog meets someone who is afraid of it, and suddenly it becomes fierce. It starts growling and showing its teeth, ready to attack, as if it was having a great time frightening its victim. Now if the person could just get rid of the fear, stand up and yell "SIT!", in most cases the dog would suddenly drop its tiger character and crawl away looking stupid. But as long as it feels the person's fear, the most innocent poodle can turn into a dangerous beast.

Many entities display similar behaviour, trying to intimidate the clearer in one way or another. It does happen that clients report things such as "the entity wants to attack you", "it says it's going to give you cancer", "it wants to hurt your children", or "when it leaves it will go straight into your dinner and then it says you'll be sorry". If the clearer reacts and gets even slightly worried, the situation can quickly get out of hand. Both he and the client are endangered. Entities are quite psychic. They will detect immediately if there is a weakness in the clearer and try to take advantage of it.

15.2 *Requirements regarding the clearer*

From what has just been said, it follows that being a clearer requires not only a particular knowledge but also certain qualities. A sufficient level of spiritual vision is needed. It won't be possible to monitor entity explorations and clearings properly unless one has a clear vision of what happens to 'the thing'. In particular, after ejecting the entity from the client, it is essential to see where it goes and if any bits and pieces are released. Perception, however, is not enough. A power is required to lift up the entity, to extirpate it from the client's energy. For this it is not sufficient to see and feel, one needs to know how to use a force.

Since releasing the entity into the atmosphere would almost certainly mean that it would go straight into a passer-by, the clearer must be in touch with a particular frequency of spiritual light, in which the entity will be processed. This implies the collaboration of certain spiritual beings. The clearer must therefore be linked to non-physical guides or angels that will assist in the process. This cannot be improvised, and requires that one has received approval from these beings to carry out the work, like a 'clearing license'. Of course these beings must be real guides or angels, not entities trying to impress you by making lots of vibration and letting bogus light rain onto your head.

Furthermore, a clearer must be cool headed and emotionally stable, able to face various threats from entities without reacting. This requires undergoing a thorough work of emotional exploration, systematically unveiling unconscious traumas and weaknesses. In the language of the Clairvision School, a clearer should be more or less 'regressed out'.

A clearer should also be quite healthy. If you are feeling fit and energetic, then clearing entities gives you even more energy. If you are weak, sick, sad, emotional, or if there is a leakage in your energy, then clearing becomes a dangerously draining exercise, and you run the risk of catching debris.

Another requirement is to have solid experience of various situations that may occur in psychic work. That's the N.B.Y. touch (Not Born Yesterday), and it helps a lot in dealing with the ninety-nine percent of entities who are only joking when they say they are going to kill you or when they invent something else.

Now what about the other one percent, the ones that are not joking? The strength of the clearer is that of the energy behind him or her. If their purpose is Truth and if their heart is one with the spiritual hierarchy that has taught them, then there is nothing to worry about. Otherwise it is preferable to have someone (physical) of a certain calibre to whom one can turn in case of major problems.

Another point that must be emphasized is that clearing entities is not only a question of level of consciousness. It requires precise technical knowledge. I have met more than one great guru in India who was incapable of clearing entities. So it is not because you have an enlightened spiritual teacher that you should assume he or she can deal with entities. For, once more, clearing entities requires a special technique. When you have a hole in a tooth, you don't ask your guru to fix it, no matter how great his enlightenment may be. You go to a dentist. Similarly, to get rid of an entity, what you need is an entity clearer.

15.3 The clearing process

After speaking so much about clearing, it would be unfair not to give any account of how I clear entities. Of course, the following indications in themselves would never make it possible for anyone to clear an entity—no need to insist any more on how dangerous it would be to improvise a clearing simply by reading a book. The following points are just mentioned so that the reader can gain a fuller understanding of the mechanisms related to our topic.

In order to remain in control of any 'fallout' of etheric and astral bits and pieces, the clearing is a strictly individual practice: one client, one clearer, and no one else in the room. No animals, and preferably no pot plants.[1]

It should happen indoors, in a pleasant and carefully chosen room (and not on a toxic earth line).

The client lies down on a mattress, eyes closed. He or she is neither hypnotised nor asked to implement any technique or particular breathing—just to stay relaxed, doing nothing but remaining aware of the entity. The clearer sits close by.

The entity should have been fully explored, so that the client can easily get in touch with it. If a client only has a vague and remote perception of the entity, if he can't feel exactly where it is, if he has not observed it as explained in the last chapter, it is better to postpone the clearing. For

---

[1] Since plants have an etheric body, entities might try to stick to them.

an unfinished clearing can generate strong resistance and is worse than no clearing at all.

It is always good to keep a candle burning when exploring or clearing entities.

The clearer utters certain sounds, to raise the level of vibration of the client's aura. The words or sounds themselves are not essential, what counts is a particular energy in the voice of the clearer. The client acts as an intermediary between the sounds and the entity. In other words, the client lets the sounds reach the entity through him.

The clearer establishes connection with his non-physical helpers, the guides or angels that will process the entity.

The clearer opens 'the Great Light', a supremely powerful quality of (non-physical) light above the head. At this stage clients often register that the energy around them is changing. Some clients even perceive so much light that afterwards they ask "Were you holding a light in front of my face?"

The clearer takes the entity out of the client and puts it into the 'Great Light'. That, one cannot explain, not for reasons of secrecy, but because there is nothing to explain. Either one can do it or one can't: it is a power and it has nothing to do with the analytical mind. At this stage, clients often see or feel the entity moving up and out of them, while powerful rearrangements are taking place in their body of energy.

The guides or angels 'heal' the entity, processing it in a way that greatly decreases its potential toxicity. This hardly takes a fraction of a second, so that as soon as the entity reaches the light it appears to change shape. For instance it often becomes clearer and lighter, it loses its gloomy or spooky aspect, or it starts smiling.

The guides take the entity to its new abode. They redirect it towards a space where it can resume an appropriate existence and no longer be a parasite. If it is a fragment, they take it to a place where it will be dissolved. Or the clearer can also explode the fragment in the Great Light, turning it into dust or undifferentiated astral light.

The client and the clearer remain motionless a few more minutes under this wonderful Great Light.

The client is gently called back, opens the eyes, and that is it. The whole process takes less than twenty minutes. At no stage is there any physical contact between the client and the clearer.

There are several other ways that can be used to clear entities, some more dangerous than others. Whatever method one may use, the same restrictions regarding the clearer remain.

The power to clear entities is not a gift that only some people have from birth. All the qualities required can be developed. Nevertheless, there is no point in writing to the Clairvision School in Sydney to ask if we can teach you how to do it. The answer would automatically be no. For even after several years of hard work, undergoing all kinds of processes, there is no guarantee that someone will reach the stage where clearings can safely be performed.

15.4 The danger of debris

Even if one does all the right things, it is not uncommon for etheric debris to be released at the time of the ejection of the entity, while it is sent from the client into the Great Light. There is nothing abnormal about this. It is like receiving droplets when splashing one's hand into water. Still it means that the clearer receives etheric (and sometimes astral) bits and pieces in his own etheric body. This release of debris also takes place when clearing cords, which is another reason why cords should be treated in the same way as entities.

I have seen a few entity clearers who were completely unaware of this mechanism, and who ended up being severely affected, even ill, because of these lumps accumulating inside their system. This fallout of debris is a subtle phenomenon. If one is not extremely perceptive or well aware of the possibility, one can easily miss it when it happens.

A simple device can be used to minimise the fallout by diverting it into something else. What do entities like? A solid majority of clients, when asked this question, immediately answer: sugar! The trick is simple: one places a lump of white sugar on a tissue close to the client before starting the clearing. Just before the ejection, the clearer tunes into the entity and presents the sugar to it, saying a few words of offering. This works remarkably well. A substantial number of the little bits that do not reach the Great Light end up in the sugar. The sugar lump must only be used once, and then disposed of—ideally buried, but not in your vegetable patch. One can also use other sweets, but preferably of a type that the clearer has not eaten recently, to avoid confusing the entity. This practice is not new. One can find such offerings as part of the rituals of all the traditions on the planet.

Once, in a house in France, I was conducting a practice called *yajña* in Sanskrit. It has to do with activating the energy in a place, and it involves a long recitation of mantras. At one stage of the *yajña*, an operation called *bhūta-śuddhi* is performed. *Bhūta* means both element and entity; *śuddhi* means purification. *Bhūta-śuddhi* is a purification of the elemental layer, and it sometimes has an effect similar to that of an entity clearing. In the *bhūta-śuddhi* practice one uses a sweet of some kind which is offered to the *bhūtas* before the clearing. Then at the end of the practice, the sweet is wrapped in a leaf and buried. That day, I was using a piece of chocolate.

In that house, there was an eighteen year old dog who had reached a more or less vegetative state. He slept from morning to night in his basket, never barking, artificially kept alive by all the vitamins his mistress fed him. As soon as I had finished the *bhūta-śuddhi* and placed the chocolate on the leaf, a most incredible thing happened. The dog suddenly jumped out of his basket and rushed over to the chocolate. To my great amazement, and before I could move, the dog had swallowed the offering to the *bhūtas*. Nobody could understand what had happened, for this dog never begged for food. Usually one had to wake him up to feed him.

Suddenly the dog was alive again! He was running through the house and even barking, as if he were ten years younger. This euphoric state lasted for three days, after which the dog gradually went back to his vegetative state. And he died, but only six months later, and probably not because of my *yajña*. This dog taught me a lesson by showing me the power of this sweet offering, which till then I had regarded as a minor and not really useful part of the practice.

In certain American Indian clearing rituals, one uses meat instead of sugar. After the practice the piece of meat is burnt, not buried. When dealing with entities that crave meat, this practice makes sense. The fact that the piece of meat is burnt afterwards is consistent with the fiery nature of meat-craving entities, which usually also like spices and alcohol, and are related to angry and aggressive tendencies.

While mentioning rituals, it may be interesting to make a remark concerning animal sacrifices. In India, I met several experts in rituals who were convinced that the animal sacrifices mentioned in ancient Sanskrit texts did not originally refer to the killing of physical animals, but to the clearing of elementals and entities. These 'animals' in Vedic texts were none other than the fragments and entities we have described in this book. The 'fire' referred to the Great Light. According to this view, it is only when the original knowledge started to be lost that texts were interpreted literally and animals were slaughtered as part of rituals.

Let us return to the important topic of the debris released at the ejection of an entity. The use of the sugar is helpful, but not enough to take care of all the fallout. Some etheric debris from the entity will probably still get attached to the clearer. It is therefore essential to recognise when

Entities

this happens. Once more, the fallout is subtle. Unless one remains vigilant and sharply aware of the possibility, it can easily remain unnoticed.

There is an interesting little sign that clearers must learn to recognise in themselves — a tiny muscular twitch that can take place anywhere in the body. It is a slight contraction of a muscle that lasts only a fraction of a second, and apparently occurs without any cause. Sometimes there is no physical twitch, just a slight etheric one.

When this happens in the minutes following the ejection of an entity, the clearer should immediately direct all his attention to that body area. There is every chance that some etheric debris has fallen on him like a dropping, and penetrated one of his energy channels. This will be confirmed by the perception of a certain density of energy in the corresponding area, an etheric vibration added onto the normal local vibration.

If identified immediately, such a dropping is not very difficult to deal with. The clearer will conduct the foreign etheric energy upwards through the channel in which it is lodged. Once it has reached the top of the head, the foreign energy is simply expelled, preferably sent into the Great Light. This doesn't require any physical movement and takes only a few minutes. It can be performed through the techniques of Channel-Release described in *Awakening the Third Eye*.[1]

If such Channel-Release can be implemented immediately and fully, it will actually have a strong cleansing effect on the clearer. Not only is the foreign etheric energy removed, but while moving it up a bottle-brush effect is generated that makes the energy channel clean and shining, much brighter than its counterpart on the other side of the body.

If such foreign energies remain undetected and are left inside, they become perverse energies. In the short term they may create headaches, fatigue and various other troubles, or they may lie dormant and remain unsuspected. In the long term, especially if more and more accumulate, they heavily undermine the clearer's physical and mental health.

15.5 Packs after the clearing

Just after a clearing, it is preferable to stop exploring for a short while. There is no purpose in trying to find out if the entity is still there or not, for the clearing creates strong rearrangements inside the client's energy, and for one or two days one can't really see anything. The client is

---

[1] One uses the technique of the 'little hands', by which one activates the movements inside the meridians, similar to intestinal peristalsis. The secret is: don't push the energy, pull it. No rubbing or physical movement of any kind should be used.

Clearing Entities

therefore instructed to stop the observation work that had been carried on until then. It is best not to worry about anything for a few days.

In order to finalise the clearing, the client will have to use a pack, or poultice, in the following days. The clearing removes the astral part of the entity (sending it into the Great Light) and also a good part of its etheric. However, it may happen that some of the etheric energy that belonged to the entity remains in the client. This is not necessarily a problem, because without the entity's astral core, this foreign etheric energy will find it much more difficult to maintain itself inside the client's body and will probably be eliminated naturally. The clearing also generates a strong activation of the client's energy, a dynamism that will get these etheric remnants on the move. Still it is preferable to help the client's system to finalise the elimination with a simple but powerful device: packs.

After experimenting with the most varied substances, some quite rare and costly, I have come to the conclusion that nothing works significantly better than cheap ordinary potatoes for drawing out energies after a clearing.

The technique is elementary. Take three or four raw potatoes. You can peel them, which will make the technique a little less messy if the skin is full of earth, but won't change much as far as the drawing power is concerned. Do not cook the potatoes. Grate them. Squeeze out the juice. Then apply the grated raw potatoes as a one centimetre thick poultice on the skin, on the area where you felt the entity before the clearing. Leave three hours, the last hour being the most important.

It is virtually impossible to get the potato pack to stay in place on the skin unless you remain somehow motionless, usually lying down on a bed or a couch. Even if you could find a clever device to fix the pack, moving would interfere with the drawing process. It is better to remain quiet for three hours, during which you can read or watch TV (but it is preferable not to fall asleep). From time to time, tune into the pack and see for yourself if you can feel anything being drained out of your body. It is often during the third hour that the draining effect is at its maximum.

Leaving the pack more than three hours is not recommended, for it could have a draining effect on the client's essential energies. For the same reason, a pack in the heart area, on the forehead or the top of the head should not be left more than one-and-a-half to two hours.

Three potato packs should be implemented after an entity clearing. For instance, one is done on the same day as the clearing, the second one two or three days later, and the last one a few days after the second (the timing is flexible). If the entity was located in the heart area, the forehead or the top of the head, only one pack should be performed.

These packs prove remarkably efficient at eliminating the etheric remnants that may be left inside the client after an entity clearing. How do they work?

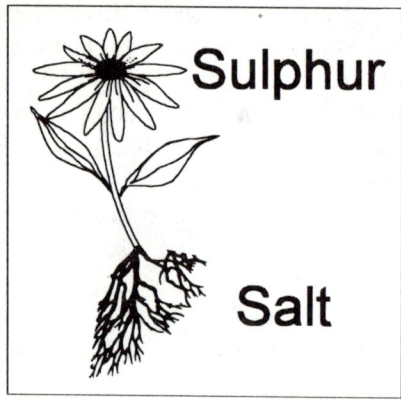

When looking at a plant, alchemists discern three processes called sulphur, salt and mercury.

Let us compare the functions of the flower and the roots. The flower radiates colours and fragrance. It is also from the flower that the pollen grains are released. The roots, on the other hand, never see the light, and draw minerals and water to the plant. So alchemists recognise two opposed groups of functions in the plant:

• The flower pole, which is related to light and colours and is giving, sending out, centrifugal. This is the sulphur process.

• The root pole, through which the plant draws in substances and gathers energies in a centripetal way. This one is called the salt process.

Mediating between sulphur and salt is the mercury process, represented in the plant by the stem and the leaves.

These three principles, sulphur, salt and mercury, are discerned by the alchemist not only in plants but in all the phenomena of nature. Alchemy being a multifaceted science, these three functions are sometimes understood in a different way. Nevertheless, an essential point is that sulphur, salt and mercury are processes, not chemical elements. In particular they have nothing to do with the physical substances called sulphur, salt or mercury.

If we go back to our potatoes, we see that even though they are not exactly roots, they still belong to the salt process which, from the point of view of alchemy, explains their capacity to draw etheric energies. Along the same line of thought, at the Clairvision School we have experimented with packs made of various types of roots and vegetables, including expensive exotic ones. We have also used alchemical ashes, a highly concentrated salt principle obtained by calcining herbs in a crucible for up to sixty hours. To be honest, apart from a few exceptional situations, we haven't found that any of these was significantly more efficient for drawing than the good old pack made of the cheapest non-organic potatoes on the market.

There is one main restriction, as far as drawing packs are concerned: **they should not be implemented before the entity has been cleared**. From all that has been said about entities, one can easily understand that they won't be eradicated with a simple pack. Even if it were possible to pack them out, where would they go? Obviously a proper clearing is needed.

Moreover, if one packs an entity before clearing it, one runs the risk of destabilising it. A few times, I have observed packs that stripped an entity of some of its etheric layer, and therefore of its anchorage. The entity then moved into another place where it was much more difficult to clear.

I remember a client who had an unclear entity in her chest, and who was finding it hard to get in touch with it. Everything was blurry and difficult to make sense of. As the client was quite psychic, I decided that she would probably be able to cope if we packed straight away instead of clearing. This was a big mistake. Twenty-four hours after doing the pack, the entity disappeared from her chest, but a terrible pain appeared in one of her teeth. The inflammation that followed required complicated dental treatment. Further ISIS sessions revealed that the entity had moved straight into the tooth area. Again this led me to the conclusion that whenever one tries to deal with an entity with anything other than a proper clearing, one exposes the client to unnecessary dangers. In conclusion, potato packs are an unexpectedly efficient and wonderfully simple remedy for getting rid of perverse energies. One can use them in various situations (not only after an entity clearing) to draw wrong energies out of the body.

The only restrictions are that, firstly, for the reason explained above, it can be dangerous to implement a drawing pack if the problem is due to an entity and not just a perverse energy. (The difference between the two is that a perverse energy is nothing more than a piece of etheric substance, while an entity has an astral part too, and therefore some mental consciousness with emotions, desires, etc.) Secondly, one should be careful not to pack too much, so one doesn't drain essential energies out of the body. In particular, one should exert caution with packs around the heart area, the forehead and the top of the head.

15.6 Vaginal bolus

The vaginal bolus is a remarkable remedy akin to packs. As far as I know, it originated with a naturopath called Dr. Christopher. It is an internal poultice made of herbs and coconut butter, and inserted into the vagina.

Entities

In the form I have prescribed, the bolus is made of equal parts of the following herbs: squaw wine, yellow dock root, comfrey root, marshmallow root, chickweed, golden seal root, mullein leaves. To this seven herb formula one adds the same global amount of slippery elm. All herbs must be finely powdered and well mixed together.

Then melt solid coconut oil butter and mix it with the herb powder until it forms a doughy paste. Shape the paste into rolls that are finger width and one inch long. Keep the rolls in the fridge, where they will solidify.

On a Monday morning insert three rolls into the vagina. The rolls, or boli, are left inside the vagina until Tuesday night. On Tuesday night, remove the boli, and douche carefully with a cup of yellow dock or burdock tea. Then immediately insert three new rolls for two days, i.e. until Thursday night.

On Thursday night, repeat exactly the same procedure as on Tuesday night, and leave the three new rolls for two days, i.e. until Saturday night.

On Saturday night, remove the boli and implement the douche procedure, but do not insert any new boli until Monday. There is nothing planned for Sunday, it is a day of rest.

On Monday start again, in exactly the same way.

Continue the process for three to six weeks, depending on your strength of character.

To retain the rolls inside the vagina, make a tampon with a piece of natural sea sponge. Sew a thread on it for easy removal. Real tampons are not necessary for the bolus, since it is not a liquid but a paste. The piece of sea sponge is sufficient to plug the vaginal opening.

Whenever possible, it is preferable to continue the bolus during the menstruation period. Otherwise start again as soon as the flow is light enough.

There is no contra-indication to sexual intercourse while implementing the vaginal bolus technique. Just remove the rolls before intercourse and insert new ones just after, and continue the same Monday, Tuesday, Thursday rhythm as if nothing had happened.

The bolus is a powerful poultice that can be used after clearing an entity in or around the vagina, the uterus or the ovaries. Apart from this, it is an interesting tool of healing for many gynaecological disorders, whether due to an entity or not. It is not rare that, at some stage during the three to six weeks, heavy discharges or foul smells occur. Afterwards, certain

Clearing Entities

problems may simply disappear, without any orthodox medical explanation.

15.7 Entities do not come back

Once properly cleared, entities do not come back. From my experience, if the entity still appears to be with the client after the clearing, it is always because it did not go, and never because it went and came back. There is a high rate of success in clearing entities because once you have mastered the technique, it is quite a mechanical process, similar to removing a pebble from a bucket.

Keeping this analogy in mind, what could be the main reasons for failure? First –and this is by far the most common– the pebble was not a pebble. The entity was not an entity, but some kind of mental image or illusion. A clearing is implemented and nothing is different afterwards, simply because there was nothing to clear. The client's problems were not due to an entity.

Second possibility: there were two or more pebbles in the bucket. The client is still found to have an entity after the clearing because there were two entities in the beginning. This can be confusing, particularly when dealing with a client who has caught several fragments from the same deceased person, for instance a close relative. A fragment is cleared, and then one finds another. The situation is confusing because the various fragments all come from the same person, and therefore they have quite a similar feeling about them. You get rid of one and you discover another one so similar that you hesitate and think maybe the first one has come back. In some extreme cases, one may have to clear more than half a dozen fragments one after the other, until nothing is left.

Third possibility: a new pebble has arrived in the bucket. Even though unusual, it can sometimes happen that a client catches a new entity in the weeks that follow a clearing. In particular, if a close relative has just died, a client may catch a fragment, have it cleared, and then catch another one in the following days or weeks. However, as we have seen, apart from temporary high risk situations, one does not catch entities easily. Clients who repeatedly catch entities are usually doing something energetically wrong, such as:

• taking drugs,

• sleeping on a noxious earth line or living in a house where the vibrations are unfit for human habitation,

Entities

• acting as a healer without really having the knowledge or the capacity to do so, and catching all sorts of perverse energies and entities from their clients,

• dabbling in magic practices.

All these possibilities remain rare, and in the vast majority of cases, no foreign presence is left after the twenty minute clearing. Of course, this does not mean that all the client's problems are solved. But removing a pebble is sometimes enough to change a destiny.

CONCLUSION

In conclusion, the main messages of this book are simple.

Entities do exist! When entering an expanded state of perception, hundreds of people who had never heard of this possibility become aware of a foreign presence attached to them as a parasite. Moreover, they describe the parasite's action in a remarkably consistent way. An 'entity syndrome' clearly appears when analysing the answers given by the clients.

In the vast majority of cases, the usual cliches about possession and exorcisms are totally irrelevant. Entities are almost always parasites and nothing more. There is nothing evil or terribly frightening about them especially knowing that they can easily be cleared in a few weeks.

There are very strict restrictions as to who can safely clear entities. Unless a number of requirements are met, the clearer can end up creating disaster. Yet it should be kept in mind that the same applies to various forms of technical knowledge, such as piloting planes or dentistry.

There are a few high risk situations for catching entities: after the death of a close relative, or for a woman after a miscarriage, an abortion or a delivery. Many problems may be avoided by having a systematic 'entity check-up' with a qualified clearer in the weeks or months following such occurrences.

Entities exist, but they are not everywhere! Moreover, one does not catch one without good reasons. So let us not develop an entity paranoia, or try to protect ourselves all the time against something which, apart from exceptional cases, cannot touch us.

An essential point is that the knowledge of entities offers rich therapeutic perspectives. Many clients who present an 'entity syndrome' have gone from therapist to therapist without getting much relief, for conventional forms of therapy are completely inefficient in their case. On the other hand, a proper clearing can give immediate results. It does not solve all the clients' problems of course, but often unlocks the situation and allows them to move towards a resolution of the trouble, whether emotional or physical.

Entities are not new. What is new is that more and more people on the planet are starting to perceive them. This will undoubtedly have important repercussions in many fields of human activity.

A

Aboriginal spirits, 102
Abortion
   and fragments, 67
   entities after -, 70
Acupuncture, 30
   Conception 17, 83
   Conception 6, 64
   Kidney 9, 84
   moxas after delivery, 82
   no - at the Full Moon, 97
Alchemical ashes, 174
Alchemy, 174
Alcohol, 10
   and catching entities, 90
Alzheimer's disease, 61
Anthroposophical Medicine, 140
Arthur, 137
Astral body
   and emotions, 44
   and sleep, 41
   as a mob, 46
   defined, 38
   reflected into the brain, 44
   shattering, 43, 48
Ayurveda
   and abortion, 68

B

Babies
   and subtle bodies, 69
bhūta
   bhūta-śuddhi, 170
   defined, 59
Birth
   and fragments, 79
   care of the mother, 82
Black Magic, 140
Breastfeeding, 83
Buddha, 97

C

Cancer, 26
Cave myth, 44
Channel-Release
   and clearing entities, 172
Characters, 46
Christopher (Dr.), 175
Chronic illnesses, 94
Clearing
   and exorcism, 138
Clearing entities
   etheric fallout, 170
   packs, 172
   restrictions, 152
   the clearer's skills, 167
   the great light, 169
Conception 6, 64
Confusion, 17
Cords, 120
   and entities, 122
   and marriage, 127
   and past lives, 130
   and relationships, 125
   and subtle bodies, 126
   defined, 121
   to a fragment, 128
Cosmic Fire, 171
Cows, 59
Cravings
   drugs, 10
   emotions, 12
   sugar and junk food, 8
   tomatoes, 9
Cremation
   no - for babies, 60
   no - for yogis, 60
   pros and cons, 55

D

Dark forces, 148, 150
Dental treatments
   and Full Moon, 97

Devas, 104
Digestion
  and entities, 63
Drugs, 10
  and catching entities, 90

E

Earth element, 101
Earth lines, 26
  and catching entities, 95
Earth-bound spirits, 106
Earthquakes, 94
Eclipses, 96
Ego
  and blood, 140
  defined, 39
  entangled in the astral body, 41
Electroshock therapy, 94
Elementals
  defined, 101
Emotional traumas
  and catching entities, 91
Entities
  and earth lines, 26
  and physical disorders, 24
  confusion, 17
  latency before manifestation, 26
  meaning of the word, 103
  secondary benefits, 20
  voices, 23
  what they want, 15
Equinoxes
  and sex, 97
Etheric body
  after death, 54
  and blood, 140
  and bones, 63
  and destabilised entities, 175
  and fragments, 55
  and menstruation, 97
  and placenta, 75
  and poltergeist, 99
  and surgery, 91
  and umbilical cord, 121
  collective awakening, 123
  dissolution after death, 48

etheric fallout, 170
of a foetus, 68
of plants, 168
thoroughly linked to the physical, 40
Excalibur, 137
Exorcism, 138

F

Faust, 35, 140
Fire, 56
Fragments, 115
  act mechanically, 138
  after the shattering, 52
  and *karma*, 111
  and past lives, 107, 119
  and samskaras, 115
  as 'spirit guides', 146
  at birth, 79
  attached to a cord, 128
  crystallisation, 49
  crystallised into poltergeist, 100
  do not necessarily look human, 98
  etheric coating, 55
  exchanged, 131
  from unborn twin, 88

G

gandharva, 104
Ghosts, 99
Gods, 104
Grail, 137
Great Light, 169
  and Vedic rituals, 171
Gui. see *Kuei*
Gulf War, 94

H

Hara, 64
Heart
  and entities, 63
Higher Ego

as opposed to fragments, 106
Hindu Gods, 104
Hun, 33

I

Illnesses, 24
Incubi, 144
Indonesia, 140
Inner alchemy
  defined, 43
ISIS
  and fragments, 107
  and Osiris, 137
  defined, 4
  opening during -, 154
  sourcing entities, 161
  spinning, 107

K

Karma, 111
Krishna's paradise, 59
Kuei
  and fragments, 106
  and sexual intercourse, 145
  defined, 35
  stories, 36

L

Leviticus, 79, 97
Ley lines. see Earth lines
Lower complex
  at death, 48
  defined, 41

M

Marriage, 127
Massage
  and catching entities, 96
Meditation
  and catching entities, 95
Menstruation, 97

Mercury, 174
Miscarriage
  as purge, 80
  of a twin, 88
Missing bits, 133
  causes, 136
  common symptoms, 134
  defined, 133
Mob, 46
Moon cycle, 96
Mother, 78
Mourning customs
  in India, 57
  in Jewish tradition, 60
Moxas, 82
Mugwort, 82

N

Narcotic Drugs, 10
Natural disasters, 94
Nature spirits, 101

O

Osiris, 137

P

Packs, 172
  restrictions, 175
  vaginal bolus, 175
Parzifal, 137
Perverse energies
  and entities, 100
  and sex, 95
  and surgery, 91
  defined, 30
  in unclean foods, 64
Physical mental consciousness, 44
piśāca, 103
pitṛ, 104
Placenta, 75
Plant elementals, 101
Plato, 44
Po, 33

Poltergeist, 99, 148
Possession, 138
Potato packs, 173
Poultices. see Packs
prāṇa, 30
Pregnancy
   arrival of the baby's soul, 67
   astral transfers, 78
Presences, 5
Purgatory, 45

Q

Qi
   defined, 30, 64
Qi Hai, 64

R

Regression
   and fragments, 107, 119
Reincarnation, 50
Relationships and cords, 125
Renan, 78
Reverberation, 44
Rock technique after delivery, 83
rākṣasa, 103

S

sūtaka
   of birth, 79
   of death, 58
Sal, 174
Schizophrenia, 61
Self, 106
   = *Shen*, 35
   and missing bits, 137
   as opposed to soul, 36
   definition of the term, 39
Separateness, 5
Seth, 137
Sex
   and catching entities, 95
   Taoist days, 96
Shattering
   described, 48
Shattering of the astral body, 43
Shen
   = Self, 35
   defined, 36
Shocks, 91
Size of entities, 6
Sleep, 41
Solstices
   and sex, 97
Sorcery, 142
Spinning, 107
Spirit Guides, 145
Spirits
   as opposed to Spirit, 106
Su Nu Jing, 145
Subpersonalities, 28
   characters, 46
Subtle bodies
   and cords, 126
   structural approach, 43
Succubi, 144
Sugar
   cravings, 8
   offering, 170
Sulphur, 174
Surgery
   and catching entities, 90
   and Full Moon, 97
   surgical theatres, 91

T

Tannhäuser, 145
Taoism
   Su Nu Jing, 145
Taoist days, 96
Tomatoes, 9, 26, 68
Transformed astral body
   building, 48
Traumas, 91

U

Umbilical cords, 121
Upper complex, 90

at death, 48
defined, 41

V

Vaginal bolus, 175
Vedic rituals, 171
Venus, 145
Voices, 23

W

Wagner, 145
Wandering souls
   practice for, 58
Womb
   fascination for, 65

X

Xie Qi, 30

Y

yajña, 170

Z

Zen enlightenment, 137